Teradyne
The First Forty Years

FREDERICK VAN VEEN

Printed in the United States of America

Design: Tepperman/Ray Associates, Inc.

Table of Contents

PART THREE
THE TURBULENT SEVENTIES

PART FOUR
THE EIGHTIES AND THE THREAT FROM THE EAST

PART FIVE
THE NINETIES. RENAISSANCE.

Foreword

Two challenges confront anyone who presumes to write a company history. The actual writing will of course carry its own challenges, but two must be faced before one puts pen to paper or opens Microsoft Word. First, is the company important enough to be worth the effort? Will anyone besides the company officers and the author's family actually read the book? Second, is the timing right? The word "history," after all, suggests a certain distance. The latest three-year-old "dot com" sensation, no matter how phenomenal, would be foolish to commission a history, while the writer who tries to capture the 100-year-old company will find that all the primary sources died long ago.

A reasonably objective analysis shows Teradyne to measure up on both counts. As it slips from one millennium to the next, Teradyne is a company of about 10,000 employees and about three billion dollars in annual sales. Many companies are bigger, and many of these did not take 40 years to get bigger. It is also fair to say that most people on the planet have never heard of Teradyne.

On the other hand...

Teradyne stands as an excellent example of the way the free-enterprise economic system works, or should work. Two young people start a company. The company grows by supplying vital products to its customers. The growth creates thousands of good jobs with good pay and benefits. Thousands of families are supported, the mortgages paid, the tuition bills covered. In the communities surrounding the company, accountants and lawyers and advertising professionals prosper. Local and Federal governments reap a harvest of tax revenues. Stockholders make a lot of money. This version of the American dream has been played out fairly often in recent years, thanks to an economic boom of epic dimensions, but relatively few companies have enjoyed a 40-year run without surrendering their identity or their values or their culture. The Teradyne of 2000, after one allows for the growth of the enterprise, is at its core remarkably similar to the Teradyne of the 1960s.

And, while it is true that most of the general public has never heard of Teradyne, it is equally true that just about everyone who works in electronics or telecommunications knows exactly who Teradyne is. It would be difficult, in fact, to find an electronics manufacturing plant anywhere — Singapore, Tokyo, Austin, Munich, Beijing — anywhere — that does not have at least one Teradyne system on its floor. Many plants have hundreds, many representing a million dollars apiece in purchase price. That brings us to the second reason why the Teradyne story is worth telling. The history of Teradyne almost exactly parallels the history of modern electronics — which is to say the history of the semiconductor. The transistor may have emerged from Bell Labs in the late 40s, but it was in the 60s that it took wing, morphing into the integrated circuit, which in turn gave us the personal computer, the

Internet, the cellular phone, and all the other solid-state essentials of modern life. As we shall see, Teradyne played a central role in making the transistor and the integrated circuit commercially viable products.

Apart from the celebration of the Company's fortieth birthday (admittedly not one of the "biggies" anniversary-wise), this history is being written at exactly the right time. It is today an enterprise of unquestioned substance, which means that it will be around tomorrow and the day after tomorrow. More to the point, each of the founders is alive and possessed of good health and sharp memory. A few of the early employees have died, but most are still around and were available for interviews. This, in short, is the right time to set it all down. In fact, it would be criminal to pass up the chance.

There are two parallel stories to be told here: One deals with the human experience of a group of people of diverse backgrounds, as they come together and work with great skill to build one of the most effective management teams the electronics industry has ever known. Most of the key players in this story traveled this journey together for 25 or 30 years, an uncommonly long time in the hurly-burly of electronics.

The other story centers on the role Teradyne played in advancing the technologies of test and connection systems. The genealogy of the Company's products is a necessary ingredient of this story, and I apologize in advance for the profusion of type numbers that inhabit the text. There was just no way to avoid it.

Something must be said in the interest of full disclosure. First, I was employed by Teradyne from 1968 through my retirement in 1995. Throughout that span I worked closely with the CEO, as his assistant and, from 1980 on, as a Vice

President reporting to him. The fact that I was close to the center of action for so many years obviously gives me a huge advantage as historian. It also gives me a certain bias. I am an unapologetic admirer of the Company and the people who ran it and who run it. So any reader looking for a "warts and all" exposé of Teradyne will be disappointed. For one thing, I never thought warts were all that interesting. Did the Company make mistakes along the way? Of course it did, and those that seem relevant to the story will be recorded. But the essence of this tale, I may as well tell you now, is that Teradyne is a great Company piloted by some of the most talented people you'll ever find anywhere. That's my story, and I'm going to stick with it.

Frederick Van Veen
Kennebunkport, Maine
February, 2001

Acknowledgments

Of the many people to whom I am indebted, three stand out: Alex d'Arbeloff and Nick DeWolf gave generously of their time and scrapbooks, and George Chamillard, who authorized the project and allowed me the kind of literary freedom most contract writers can only dream about. Since Teradyne was underwriting this book, it could have insisted on editorial control and approval of every adjective. That it didn't is one more measure of the absolute integrity of the Company.

I am also grateful to the following people for spending the time to answer my questions and for sharing their memories with me: Brian Amero, Larry Apfelbaum, Russ Ashdown, Dick Bailey, Rick Bourne, Mike Bradley, Byron Brooks, Frank Burge, John Casey, Nick Cedrone, George d'Arbeloff, Faith DeBourgnecht, Rick Dyck, Jim Hahn, Joe Hendriks, Lucretia DeWolf Hosmer, Mark Jagiela, Peter Kelley, Peter Konde, Joe Lassiter, Marc Levine, Dick Lucey, Bill McCandless, Jim Mulady, Tom Neavitt, Tom Newman, Dennis O'Connell, Stu Osattin, John Petrolini, Jim Prestridge, Owen Robbins, George Rotsky, Bill Russell, Michael Schmidt, Rick Schneider, Dave Sulman, Kemon Taschioglou, Norm Turner, and Hap Walker. In addition the following Teradyne employees were immensely helpful in collecting much of the data, photographs, and other material that found its way into the book: Mary Ellen O'Connor, Craig Pynn, Joan Burns, and Andy Porter.

Finally, the fact that the text is not teeming with misspellings and misplaced commas is directly attributable to the expert proofreading of my wife, Jill, and of Kit Rosenbauer.

References

The primary source materials for this book were Teradyne publications, including the annual reports, as well as the Company's correspondence files, the author's personal notes covering the period, interviews with many employees and former employees, and material from the private files of Alex d'Arbeloff and Nick DeWolf.

Those wishing to read a short version of Teradyne's history can do no better than Professor Joseph Lassiter's "Teradyne, Inc. Nothing Ventured, Nothing Gained," published by the Harvard Business School in 1998.

Prologue

Teradyne exists because the lives of two extraordinary young men intersected in 1960. Then in their early thirties, the two had gone their separate ways since their student days at MIT, where they knew each other only casually. Chance brought them together again at a fortuitous moment, and their decision to start a company struck a spark that eventually touched thousands of other lives, almost always for the better. In fact, if one includes in the total the contributions their company made to the semiconductor revolution, as indeed one should, the total population of those affected runs into many millions.

Many companies have been founded by pairs of gifted entrepreneurs. In electronics, one thinks first of William Hewlett and David Packard, assembling their oscillators in a garage in Palo Alto. It is highly doubtful that either Hewlett or Packard could have built so great a company without the other. The same can be said of the Teradyne founders. Their talents were entirely complementary, each bringing to the enterprise talents the other lacked. Both possessed the oversized egos that usually go with the territory, yet their skill sets overlapped so little that there was no sense of competition to interfere with the pursuit of their primary objective.

The two young founders did not name their venture after themselves, and for that reason, no matter how much their Company has accomplished or will accomplish, their names will never enjoy the recognition that rightfully belongs to Hewlett and Packard. In fact, many newer employees of today's Teradyne may not even know who founded the Company. Many of them were not even born in 1960, when it all began. Moreover, one of the founders bowed out around 1970, and, despite their strong personalities and egos, both were rather inept (or uninterested) in the art of self-promotion.

This book will attempt to set things right: The two founders of Teradyne were enormously talented human beings, who worked incredibly hard to fashion a solid company out of their shared vision. To some extent, one can identify the talent and the work ethic as trace elements handed down by their forebears, so the story of Teradyne begins with the fascinating sagas of the families DeWolf and d'Arbeloff.

PART ONE

Nicholas and Alexander

1

• • •

Rebel with a Cause

Nicholas DeWolf was born in Philadelphia July 12, 1928, the first child of Maurice Mortimer DeWolf and Marian Watts DeWolf. Marian, a Main Line debutante, was a great-granddaughter of the famed financier Jay Cooke. At the time of the U.S. Civil War, when the Union's weak credit standing made it impossible for President Lincoln to fund the campaign by conventional means, Treasury Secretary Salmon Chase turned to Cooke for help, appointing him the exclusive distributor of Government bonds, at a gross commission of a quarter of one percent of the total amount raised. Eventually Cooke's monopoly would attract political controversy, but two years on the inside track was enough to generate $220,000 in profit and establish the family fortune. After the War, Jay Cooke and Company became the largest dealer in government bonds, with offices in Philadelphia, New York, and London. Cooke's skill at making money out of money was shared by a son-in-law, Charles Barney, who was a founder of the investment banking firm of Smith, Barney — and Marian DeWolf's grandfather.

Marian was thus a member of a family of substance, and she was well educated (Smith, '23), beautiful, and talented enough to entertain notions of a stage career. Maurice's roots were more modest — his father was a Grand Rapids doctor — but he was an Annapolis ('24) graduate, handsome in his Navy whites, and equipped with a sense of style that would serve him well throughout his career. The young couple began their married life in Philadelphia but moved, when Nick was one year old, to New York City, when Maurice decided that a career in coffee and sugar brokerage might be more exciting than anything the peace-time Navy had to offer. At about the time of the move, Nick's parents gave him a sister, christened Lucretia and called Lucky.

From his earliest years, Nick displayed an interest in finding out how things worked, if necessary dismantling or destroying them in the process. One episode, which stands as Nick's earliest known fling at testing, occurred during a visit to grandfather Watts's beach house in Bay Head, New Jersey, when he was about six. "Nick had a toy fire engine with a hose that squirted water," his sister recalls, "and Nick set a field on fire so that he could see how well the fire engine worked."

Home life, as Nick remembers it, was suffocatingly Victorian, with one exception: Instead of the Queen's English, the young DeWolfs were required to speak French at the dinner table. Nick bridled at such attempts to force-feed *savoir faire* and resisted the whole idea of becoming a proper young gentleman. No matter. In New York, he was enrolled in Trinity School, at 139 West 91st Street, to which he commuted by bus from the family's 72d Street apartment. Trinity, founded in 1709 (!) "under the auspices of the English Society of the Gospel in Foreign Parts," had a clearly defined mission "to send its boys into a larger world with a desire and capacity for leadership, and a strong sense of responsibility to society." What its mission did not include was breaking new ground. "Untested theories," the School's catalog huffed, "are not made the basis for experimentation."

The record shows that Nick was an outstanding student at Trinity. There were glowing reports from his teachers in English, art, music, French, and arithmetic, and this over-all assessment from the Headmaster:

June 19, 1940

Dear Mr. and Mrs. DeWolf

It is a pleasure to send you this final report for Nick, since his average for the year is 91.8.

He is one of the most intelligent boys in the class and is outstanding in everything he does.

It is a real pleasure to have him here and I am sure his fine work will continue next year in Form I.

Sincerely yours,
Matthew Dunn

At the time of this letter, war had broken out in Europe, and it was about to engulf the DeWolf family. In 1936 Maurice DeWolf, tiring of the brokerage business and sensing a war in the offing, had reactivated his Navy commission and returned to Annapolis, where he became an instructor in seamanship and navigation, as well as Commandant of the Academy's sailing fleet. He was joined by the rest of his family in early 1941, and Nick and his sister were soon sailing around Chesapeake Bay with their father, in the Academy's 30-foot yawls. Nick's education was resumed at the Severn School, an hour's train ride from the DeWolf's Annapolis home, and once again his scholarship was applauded by his teachers. At home, he earned his applause by finding new ways to entertain his parents and their guests. A pet chicken became part of the act, and the young experimenter learned how to hypnotize it and cause it to lie down on the living-room carpet.

He also found new ways to apply his zest for tinkering. Model airplanes were popular at the time, and Nick built them by the dozen, carefully gluing the struts in place, sanding and painting the fuselages and wings, and applying the insignia decals. He was also fascinated by the opportunities for mischief inherent in jet propulsion, and on one occasion found some saltpeter caps of a type used as railway warning devices, loaded the powder from these into empty cartridge shells scavenged from the Navy's firing range, added a toilet-paper wick, installed the whole business in a little silver toy car, and proceeded to blow it up, showering his young sister with shrapnel.

The Japanese attack on Pearl Harbor devastated the Annapolis community, and the shock waves from the "date which will live in infamy" hit the DeWolf household especially hard, as two of their closest Annapolis neighbors were killed in the bombing. The DeWolf children heard the news and concluded that it was just a matter of time before their own father was added to the casualty list. The news that Maurice was assigned to the battleship Massachusetts confirmed the family's worst fears, for everyone "knew" that when battleships went down, they went down with all hands. The Massachusetts, as a matter of fact, did not go down, but it did see extensive action in both the Atlantic and Pacific theaters. It earned its place in naval history in November, 1942, when it supported an Allied invasion of North Africa (Operation TORCH) and destroyed the Vichy-French battleship Jean Bart at Casablanca.[1]

The war took Maurice DeWolf away from his children for several years, and it also took Marian away, since she made heroic efforts to be near the battleship's ports of call. While she traveled, Nick boarded with an Annapolis family of whom he remembers little. Today he recalls his mother with affection, his father with antipathy. Yet, whether or not Maurice failed his children, he certainly did not fail his country, serving with distinction on the Massachusetts and, after the war, as Naval Attaché at the American Embassy in Paris. Captain and Mrs. DeWolf enjoyed a life rich in adventure. They were together in China when the Communists took over, fleeing the country with the forces of General Chiang Kai-shek. They lived their last years in Europe, seeing little of their children.

As war broke out, Nick was sent to the Pomfret School, in Pomfret, Connecticut, a way station for privileged young men en route to Yale or Princeton. There Nick was surrounded by demons from the start. Like most prep schools, Pomfret placed great store on sports, with its football, hockey, baseball, and basketball leagues competing against teams from St. George's, Providence Country Day, and Dean Academy. There was crew for those so inclined, and, for those with marginal athletic ability, there was always the intramural program, with the Ionians

[1]Captain DeWolf was Damage Control Officer on the Massachusetts in both the Atlantic and Pacific theaters, and he apparently did his job well: "Big Mame" was in the thick of action at Tarawa, Kwajalein, Truk, Saipan, Okinawa, Iwo Jima, and many of the other Pacific hot spots, yet she survived all this plus two typhoons without losing a single crew member! Today she sits proudly at anchor in Fall River, Massachusetts.

facing off against the Achaeans. Enough opportunities, in short, for everyone on campus… except Nick.

Gangly, bespectacled, younger than most of his classmates, and more interested in reading Tom Swift than in shooting baskets, Nick quickly assumed the role of outsider. Moreover, the aura of privilege that hung over the ivied campus reinforced his anti-establishment feelings. He recalls Pomfret as a thoroughly miserable experience, in which he had "no family, no love, no friendship." This is surely an overstatement, since his Yearbook, in which he is labeled "The Mad Scientist," reveals that he was a member of the glee club, the choir, and the dramatic club (acting in "H.M.S. Pinafore"). He was photographer for the Class Book and a member of the Banquet and Banner Night Committees in his senior year. But no sports, and this did mark him a rare bird at Pomfret. (Today, although he remains steadfast in his sour opinion of the school as it then was, he recalls individual Pomfret classmates with warmth. It was clearly not the students that turned him off, but the world he felt the school represented.)

Nick was an average student at Pomfret, with excellent grades in Physics, Chemistry, and "Radio," and middling grades in English, German, and History. (French, he recalls with relish, was considered effete at Pomfret during the war years, while German was quite acceptable.) His Yearbook, summarizing the highlights of his senior year at Pomfret, offers the following glimpse of "The Mad Scientist" at work:

> "Unable to light his incendiary bomb, Nick DeWolf surprised us all by building an entire broadcast station unknown to Mr. Huntley and was only dissuaded from a nationwide broadcast of "Halfway Down the Street" at the last moment."

Most of Nick's Pomfret classmates would migrate to the Ivy League, as programmed, but Nick saw Princeton as Pomfret writ large, and his scientific bent led him to apply to MIT. A science teacher at Pomfret put in a good word for him, and he was well prepared for the entrance exams. So, upon graduation from Pomfret, Nick headed for Cambridge. The realities of war and the draft prompted many colleges, MIT included, to forego summer vacations in order to speed students through as quickly as possible. Nick therefore entered MIT in July of 1945.

Life looked brighter the minute he arrived at the MIT campus. At last, he was a square peg in a square hole. At MIT, success or failure depended not on genealogy but on merit, which was defined not by how well one threw a football but by how well one did academically. Here, being a "geek" instead of an athlete was no disgrace, but a badge of honor. He was invited to join Sigma Alpha Epsilon, lived at the SAE house on Beacon Street, and aimed for a degree in electronics engineering. All the evidence suggests that he was as popular at MIT as he was unpopular (or thinks he was) at Pomfret. Now under no pressure whatsoever to participate in sports, Nick joined MIT's sailing team and was quite good at it, having mastered the finer points on Chesapeake Bay with his father.

He fell in with a congenial crowd and filled his off-hours with the collegiate pursuits of the day — playing cards, going to parties, chasing girls. Poker particularly appealed to him, because he found that the game yielded to his brand of strategic thinking. One of his poker stories is worth the telling:

> "After playing a lot of poker with these guys, I figured out that the secret of poker was alcohol. Everyone drank like crazy at the beginning of the

> night and was stoned by 10 PM. But I knew that all the real money changed hands at 2 AM. So I quit drinking at 11 and was sober by 1:30. I did this every night, waiting for "the big hand." Finally it happened. The pot was sky high. Guys were coming out of their sacks to watch it. The players were all hard-drinking veterans.[1] There was probably two or three thousand dollars in the pot. The president of the house was looking on. I looked at my cards and said "pot," which meant I was betting an amount equal to what was in the pot at the time. This veteran, the only other player still in the game, says "three pot," which meant he was tripling the bet, but he didn't have enough credit to do that, so he bet his car. Now, a car was a big deal in those days. A car meant you could drive to Wellesley College and make out. Anyway, the vet had four queens, I had four kings, and I had a Ford convertible."

The car, which Nick called Gladys, is remembered by Kemon Taschioglou, a classmate and later a Teradyne employee. "One time three of us went off on a jaunt together in Gladys, to western New England. Nick would pull the horn out of its socket in the steering wheel and steer the car with it. He was eccentric and always very amusing."

With the war still casting a shadow, and the draft still in force, ROTC (Reserve Officers Training Corps) was a popular elective at most colleges, including MIT, and Nick signed up enthusiastically. Because the ROTC ranks were arranged

[1]The reference to veterans places this story in Nick's sophomore or junior year, when the GI Bill brought hordes of war veterans to the campuses of America.

alphabetically, DeWolf found himself adjacent to a classmate named d'Arbeloff. The two traveled in different circles at MIT, and the ROTC connection did not foster a friendship, although they remained vaguely aware of each other in the years that followed. Their closest brush at MIT was at ROTC summer camp, at Aberdeen, Maryland. (They served in the Ordnance Corps.) While at Aberdeen, Nick asked for extra time on the firing range, explaining to his sergeant that since the whole idea in the military was to kill people, he wanted to be good at it. Tongue in cheek or not, he got his wish, won the Expert badge, and wore it proudly.

Nick graduated from MIT in the summer of 1948 and accepted a job offer from General Electric. His training program at GE brought him first to Burlington, Vermont, a plant that assembled aircraft machine guns, then to the Pittsfield (MA) power transformer plant, and finally to GE's Electronics Park, in Syracuse, where the Company made television tuners. This turned out to be a pivotal assignment, as it brought him face to face with the crystal diode. The crystal diode was quirky, like Nick, and it was love at first sight.

At the time, GE, like most other companies involved in consumer electronics (and indeed, like most Americans), was fixated on television. It had already become clear that twelve VHF channels were not enough to satisfy the demand for broadcast outlets, and the first forays into UHF, a mysterious territory north of channel 13, had begun. The technical difficulties were considerable. Higher channels meant higher frequencies, where only very expensive vacuum tubes could operate. The solution lay in the use of the semiconductor diode as a "mixer" to help convert the UHF frequencies into the lower frequencies the standard TV receiver could handle. The solid-state diode could be made inexpensively, it didn't draw much power, and the good ones worked reliably.

The trick, as GE and others soon found out, lay in knowing which diodes were good enough and which weren't. That meant measuring a characteristic called noise figure. At UHF, it was all too easy for a circuit to generate electrical noise, and a noisy mixer circuit could distort or even wipe out the television signal. DeWolf therefore set out to develop instruments that would measure all the parameters that contributed to noise figure. Within two years, he was GE's diode guru, and he began giving papers on the subject at technical conferences and writing articles such as the definitive "Testing UHF-TV Mixer Crystals," which appeared in the April, 1953 issue of *Electronics* magazine.

DeWolf's new-found celebrity offered GE a triple payoff: The Company could sell diodes (like its immensely successful 1N72) to television makers like Philco on the strength of quality verified by superior testing. Its mastery of diode production enhanced its reputation in the embryonic semiconductor business. And GE was even able to market a diode tester designed by DeWolf. The instrument, designed for laboratory use, would not set the world on fire, but DeWolf's interest in testing extended far beyond the laboratory, and an internal GE paper authored by him in 1952 covers many of the themes around which he would ultimately create a major electronics company — and, for that matter, an entire industry. An excerpt:

The Design of Factory Production Test Equipment

1. It must be usable.
2. It must be safe to use, and almost impossible of causing injury.

3. It must be reliable and use good components and circuits.
4. It should be simple, to ease maintenance.
5. It must be easily calibrated with simple gear.
6. It must operate in a wide range of supply voltages.
7. It must operate in a wide range of temperatures.
8. It must be reproducible. If several sets are built, they MUST read the same thing, even if it involves a loss in overall absolute accuracy.
9. It must be simple to maintain and accessible.
10. The operating portion of the gear at the operator's station must be small to prevent overcrowding.
11. It must test only the desired characteristics and not be responsive to other parameters, or varying quality may result as product changes are made.
12. It must be accurate to at least a 1% yield rate sensitivity.
13. It must be stable. A good criterion is a 1% change in yield rate over a 2-day period without calibration.
14. It must be capable of a complete test in a short period of time, or the testing expense will increase.
15. It must require a minimum of thought on the part of the operator.
16. It must not fatigue the operator, or mistakes will occur.
17. It should be reasonably neat in appearance. This instills confidence in the equipment.
18. It should be flexible enough to allow for specification changes that inevitably occur.
19. It must favor the rejection of bad units over their acceptance, in respect to predictable drifts and failure. This applies particularly to automatic test equipment — fail safe!
20. It should be reasonably quiet and cool and not damage tested units.

21. Remember the operator, who is 2/3 of the equipment.
22. It must give repeatable results.

The list is historically important as an early insight into DeWolf's philosophy of test equipment. (Note, in commandment 19, the reference to "automatic test equipment," a term that would not come into vogue until 10 years later.) One supposes that some GE old-timers, reading this ukase handed down by a cocky 24-year-old engineer, shook their heads in derision. But GE did not become GE by scoffing at sound thinking, and publication of this list simply added to Nick's luster in Syracuse.

DeWolf was generally happy at GE, to which he gives full credit for launching his career and supporting him in his early efforts. But the thought of spending a lifetime working for a large, highly structured, basically conservative company — one, moreover, that took its traditions very seriously — was not an appealing prospect to the young iconoclast. So, after four years at GE, he began putting out word that he was available.

It didn't take long for an offer to arrive. At a conference on solid-state electronics in New Jersey he met David Bakalar, an ex-Bell Labs engineer who, with his brother Leo, had just formed a new semiconductor company in the Boston area. Nick was offered the chance to join the venture, called Transitron, as its third employee and its Chief Engineer. The chance to help create a new company in an exciting field was irresistible, and on September 5, 1952, Nick sent a cordial letter of resignation to GE.

Transitron caught the semiconductor wave at exactly the right moment, and within a few years it was the hottest of the new breed of semiconductor manufacturers. Its considerable early success was based on an innovative gold-bonded diode. Later there would be transistors, but it was the diode that put Transitron on the semiconductor map and fueled its growth, and it was DeWolf's genius at developing automatic diode testers that helped Transitron keep its costs low and its profits high. Its meteoric rise was chronicled in a 1959 Fortune article,[1] which pegged its sales at $42 million, making it the world's second largest semiconductor maker, after Texas Instruments. There is no question that DeWolf's engineering talents played a major role in the Transitron success story. The Fortune piece noted that he was also considered "Transitron's best salesman."

This observation was perceptive. Although DeWolf's core strength was his overpowering intellect, it was his enthusiasm and his ability to project this enthusiasm that most people remember. There were other ace technologists at the time, but none could match DeWolf's charisma. He was tall, reed-thin, with russet-colored hair, a smile that wouldn't quit, and a high-pitched voice — not squeaky, but modulated and expressive. When he spoke, he would punctuate his comments with appropriate flourishes of his hands and arms. He seemed to be a good listener, but he was always in command of every conversation. And he laughed a great deal, whether the jokes were good or bad. He was impossible to ignore, or to dislike.

At the time DeWolf joined the infant Transitron, the transistor was little more than a curiosity, used in hearing aids and some radios. (It is worth noting that worldwide semiconductor sales were only $20 million in 1952.) In general, the device was prized purely for its potential in consumer electronics — the television receivers, for instance, that were the focus of

[1] "The Company That Started with a Gold Whisker," William B. Harris, *Fortune*, August 1959.

DeWolf's work at GE. DeWolf was among the first to see that there was bigger game available: computers based on the use of semiconductors, not as amplifiers, but as switches. David Bakalar shared the vision, and Transitron started aiming its engineering guns away from consumer devices and toward the industrial market, winning a milestone order for 25,000 diodes from Remington Rand's Univac Division in 1953. At about the same time, the Navy turned to Transitron to replace unreliable point-contact diodes in its computers with the gold-bonded devices that were Transitron's specialty.

Transitron during the DeWolf years was a fast-growing, exciting, and important company. Now all but forgotten, it deserves its own chapter in the history of the early semiconductor industry. Its diodes and transistors broke new ground in several areas, its laboratories produced new types of devices (including the binistor, a switching transistor created by DeWolf and now popularly known as a silicon-controlled rectifier), and its factories pioneered many of the production techniques which later became standard — including test techniques which would be more fully developed by Teradyne.

DeWolf's passion for semiconductors was great, but not great enough to crowd out the other pleasures of life. He was a gregarious young man, and since he was the life of any party, he was invited to many. Some of these were organized by the Sitzenjibbers, a group of Boston-area college graduates nominally bound together by an interest in sailing and skiing but in reality by a love of partying. Above all else, of course, a Sitzenjibbers' party served as a matchmaking bazaar, and at one such event Nick met a lively brunette known as Maggie (Margaret) Lemle. She was a southerner, the daughter of Louis and Amelia Lemle, he of the New Orleans law firm Lemle and

Kelleher. The couple were married in 1958 and made their home on Beacon Hill, in a Pinckney Street town house.

DeWolf and his fellow engineers and managers found creative freedom and a high-adrenaline environment at Transitron. What they did not find was equity participation. The brothers Bakalar owned 100 percent of the Company throughout its first seven years — years in which its sales and profits skyrocketed. There was no public market for the stock, no stock options. Nick may have been lionized by Fortune and widely recognized as a *wunderkind*, but there was no opportunity to share in the real wealth his talent was helping to create. Eventually the Bakalars did take Transitron public, but then it was too late; engineers who for years had been driven to keep their costs down now could see, in the offering prospectus, just how profitable their Company had been all along. Suddenly, other semiconductor companies found that the talent they desperately needed was ripe for the picking at Transitron.

Nick DeWolf said good-by to Transitron in early 1960, less than a year after the Fortune article had tabbed him one of that Company's most valuable employees. At the time he was a 32-year old engineer with a solid reputation and a card file full of ideas. Many of the ideas were brilliant, but as he reviewed them he came to a fundamental decision: Success — *real* success — lay in choosing, not the right invention, but the right field.

What, in 1960, was "the right field"? Semiconductors? He could create his own Transitron, avoiding the mistakes of the Bakalars. It was tempting, but in his view it was too late to begin building a semiconductor company.[1] But it was not too

[1]He was wrong, of course, and it is interesting to speculate how a Nick DeWolf might have altered the semiconductor landscape in the next few decades. He was among the first to believe in the transistor's potential as a low-power switch, and had Teradyne been founded as a semiconductor company, there is no telling where DeWolf's zeal and genius could have carried it.

late to be a true pioneer in testing. Given his work in developing in-house test equipment at GE and Transitron and his strong convictions on the subject of testing, there was really only one business that made perfect sense for DeWolf, and he began thinking about the kind of company he would create. One thing was sure: He could not do it alone. He was realistic about his limitations, and, notwithstanding his reputation as "Transitron's best salesman," he knew that his lack of business experience was a handicap. Fresh in his mind, also, was the example of the Bakalar brothers. Just as David the scientist needed brother Leo, the money man, Nick, too, would need a partner.

2

• • •

The D'Arbeloff Saga

Manhattan, two days before Christmas, 1936: Eight-year-old Nick DeWolf waits in the cold for the bus that will take him to Trinity School. Christmas vacation will begin tomorrow, and that's a consolation, but not much: Father has gone back into the Navy, to Annapolis, and Nick feels deserted.

As Nick waits for his bus, the *Europa*, pride of Germany's Hapag-Lloyd line, cuts through the morning mist of the outer New York harbor en route to its Hudson River berth. On deck stand 38-year-old Vladimir d'Arbeloff, his 33-year-old wife Catherine, and their two sons, Alexander and Dimitri, nine and seven. The Statue of Liberty comes into view, torch lifted to welcome the liner and her passengers. But the d'Arbeloffs are not the wretched refuse of a teeming shore. They are a well-off family who have left their comfortable home in Paris because of the growing threat of war in Europe. To the elder d'Arbeloffs, it is an old story: Two decades earlier, they had fled their homeland at the onset of the Russian revolution.

Catherine's world had been turned upside down by the uprising. She was the daughter of the Baron Tiepolt, a flag officer in the Czarist Navy, with a distinguished career and family ties to Russian royalty. Not surprisingly, then, he was arrested by the Reds and whisked off to prison. For an extended period after his arrest his wife (the Baroness Elizabeth Tiepolt), young Catherine, her brother Nicholas, and two daughters from the Baron's first marriage all lived in Petrograd (formerly St. Petersburg, later Leningrad) in greatly reduced circumstances, reminiscent of a memorable scene in the movie *Doctor Zhivago*. Eventually, the Baron was freed from prison on the curious technicality that his ancestry was not authentically Russian, but Luxemborgian (!). The Baron then won exit visas for himself, his wife, and daughter Catherine, and the three were off to France. Son Nicholas headed east with the White army, eventually winding up in Shanghai, and Catherine's half sisters remained in Russia.

Vladimir d'Arbeloff was a native of Georgia, living in Baku at the time of the revolution. Family lore refers to him as the son of Alexander Prince Arbelov. ("To be a prince in Georgia," says a dubious Alex today, "you had to have at least *three* sheep.") Princes or not, Vladimir and his older brother Sasha (Alexander) were survivors, and when the revolution came the two beat a retreat from their homeland, first to Berlin, then to Paris, where the Russian émigré community was well established.

In the Caucasus, the family name had been Arbelov (probably a Russianized version of the Georgian "Arbeliani"). Once in France, the prefix "d" was added, following the French method of denoting a family of noble (or allegedly noble) origin.

In Paris, the two brothers found their way in and out of a variety of ventures, succeeding on wits and style. When asked how his father earned a living, Alex simply says, "He was a promoter," and the evidence suggests that he was a very good one. Sasha was at least as adroit. At one point he threw in with the noted French movie director Abel Gance and actually helped Gance produce his masterpiece, "Napoleon."

Of Vladimir's various "promotions," the most serendipitous was a turn at dressmaking, for that is how he met Catherine Tiepolt, as she practiced her craft of embroidery in various fashion workshops. The two met in 1925 and were married in Paris the day after Christmas, 1926, in a double ceremony that also united brother Sasha and his Parisian bride Blanche Augustine. Almost exactly a year later, on December 21, 1927, Catherine gave birth to her first child. He was named Alexander, the hero of our story. Two years later Catherine would bear a second son, Dimitri, called "Dee." Sasha and Blanche would keep pace, producing two daughters, Anne and Natalie.

Life in Paris was pleasant enough in the thirties, and the d'Arbeloffs were able to live well. The apartments at 74 Avenue Kleber and later at 10 Boulevard Suchet were comfortable, a testimony to Vladimir's success as a businessman. His various ventures would also require him to travel abroad, and several trips to the United States provided contacts that would one day be useful.

Paris was also a congenial home for the Baron and Baroness Tiepolt, restored by the Parisian Russians to a position of prominence, the Baron becoming a deacon at the former Russian embassy church on Rue Darue. The extended d'Arbeloff family thus included Vlad and Sasha, their wives and children, and Catherine's parents — plus an elderly Russian couple named Avram and Clara Promischlansky, possibly related to the d'Arbeloffs but in any event de facto family members. These two were known to young Alex and Dee as "Dedushka" and "Babushka." Little is known of their backgrounds, other than the fact that Avram at one time was the publisher of a newspaper in Baku.

In March, 1936, Hitler renounced the Treaty of Locarno and seized the Rhineland. In May, Italy marched on Addis Ababa. In July, civil war broke out in Spain. Much of the world took these events in stride, and despite Hitler's growing belligerence relatively few people believed that war was imminent. But those whose antennae had been sensitized by the events of 1917 knew that a European war was in the making. On December 17, 1936. therefore, Vlad and Catherine d'Arbeloff, along with their two sons, Catherine's parents, and the ever-present Dedushka and Babushka boarded the *Europa* at Cherbourg. They arrived in New York City, not as immigrants, but as travelers in transit, en route to South America, where they would join Sasha and his family. The d'Arbeloff's sojourn in New York City lasted about three months — long enough for Vladimir to conduct some undefined business,[1] and then it was off by steamer to Buenos Aires, thence by river boat up the muddy Paraguay River to Asuncion, Paraguay, where Sasha and his family were on hand to welcome them.

Why Paraguay? Because it was to be the scene of yet another d'Arbeloff "promotion": the construction of a highway from the capital, Asuncion, eastward through the jungle to Brazil. The fact that neither Vlad nor Sasha knew the slightest thing about road-building was neither here nor there.

Home in Paraguay was a plantation in the dusty town of San Antonio, about ten miles downstream from Asuncion.

[1]A reasonable guess is that Vladimir used the time to convert some U.S. assets into capital needed for his South American venture.

The Baron and Baroness moved into the existing ranch house, and a new, larger house was built for the rest of the family. For a full year this would be home to the d'Arbeloffs, who now added Spanish to the French and Russian languages they brought with them. Whatever the language, Alex was always an inveterate reader, and life on the *estancia* offered a wonderful opportunity to devour books — especially since the circumstances did not allow any formal schooling. It also gave the family a ringside seat to the revolutions that were part of the fabric of the country. In one instance — probably related to the overthrow of Colonel Rafael Franco by Felix Paiva — the rebel forces "requisitioned" the rowboat the d'Arbeloffs kept on the river bank and commissioned it part of their navy.

As time passed, the highway project fell victim to the ever-changing politics of Paraguay,[1] and by late summer 1938 it was clear that this was not to be one of the d'Arbeloffs' successful promotions. Plans were then laid to move the entire family to its next destination: the United States.

The logistics of the move were complicated. First Vlad and Sasha flew to New York to prepare the way. The wives and children, plus Babushka (Dedushka had died) endured the three-day river voyage downstream to Buenos Aires, then transferred to a coastal steamer to Rio de Janeiro, where they spent a few months before continuing on to New York, in the fall of 1938, in time to enroll the children in school. In New York, the d'Arbeloffs once again set about acclimating themselves to a new homeland. The resettlement was not without separation and regret: The Baron and the Baroness had decided to remain

[1]Ironically, only a year later, the Export-Import Bank in Washington granted Paraguay's new President a $3.5 million loan specifically for the construction of a highway to Brazil. This, too, would founder on the political shoals, and the highway would not be built for many years. Today it connects the capital city of Asuncion with the border town of Puerto Presidente Stroessner.

in Paraguay, which, revolutions or no, they judged a safer bet than the war-threatened continents to the north. The Baron would remain in Paraguay until his death in 1947, after which the Baroness would spend her remaining years in the United States, near her daughter and her grandchildren.

Alex was now 11 years old, a bright enough lad with three languages in hand — not including English. He was thus assigned to the fourth grade at a Friends' (Quaker) school in New York City, where he was thrown in with much younger children. Of the experience, he remembers only being violently ill in class one day, causing the entire student body to be evacuated, and watching his classmates file out while he lay helpless on his back, unable to explain his plight because he couldn't speak the language.

This episode occurred in late 1938, a period when, interestingly enough, young Nick DeWolf would have been equally miserable nearby, at New York's Trinity School. But soon enough, both d'Arbeloff and DeWolf would escape New York, Nick to Annapolis and Alex to, of all places, Hollywood, California.

(The reader at this point may be forgiven if he or she feels that this narrative has crossed the line into fiction. Paris? Paraguay? Hollywood? It is too much.)

In the summer of 1939 Vlad and Catherine decided to travel in their newly adopted land, *sans* children. The d'Arbeloff family therefore crossed the country to Los Angeles, where they deposited young Alex and Dee with a distant relative named Jacques Grinieff, in a house just off Sunset Boulevard. First, however, there was a quick trip over the Mexican border and back in order to establish the d'Arbeloff family as "immigrants" and thus make them eligible for eventual naturalization. (Up to that point they had been in the

United States under a tourist visa.) This common procedure, arranged by an immigration expert hired by Vladimir, saw the family drive across the border to Tijuana, then back to the San Ysidro port of entry, where they obtained the essential immigration papers. Shortly thereafter, Vlad and Catherine began their sightseeing trip, and Alex and Dee began their adventure in Hollywood with Jacques Grinieff.

Grinieff, along with Sasha d'Arbeloff, had been a co-producer of Abel Gance's "Napoleon," and he had since emigrated to the United States and joined the Hollywood expatriate scene. There he worked in films as a producer, in the company of such celebrities as fellow producer Sam Bronston, remembered by Alex for his gruff, "Can't you keep those kids quiet?" The Grinieff house, which bore the distinction of having once been the home of the actress/diva Grace Moore, afforded the two young d'Arbeloffs a crash course in American culture, or what passed for culture in 1939 Hollywood.[1] Soon Alex was speaking English well enough to merit a double promotion, from the fourth grade in New York to the sixth grade at the Third Street School in Los Angeles.

The Hollywood chapter of Alex's life lasted less than a year, for in the summer of 1940 the two boys were retrieved by their parents and settled into the family's new apartment in Kew Gardens, Queens. Alex entered the seventh grade at P.S. 99, and a year later he was enrolled at Brooklyn Technical High School.

[1]It is worth noting in passing that the year the young d'Arbeloffs spent in Hollywood is judged by many to have been the greatest year in film-making history, producing *Gone With the Wind, The Hunchback of Notre Dame, Ninotchka, Goodbye Mr. Chips, The Wizard of Oz, Gunga Din, Wuthering Heights,* and many other classics. But to 12-year-old Alex and his 10-year-old brother, the cinema highlight of the year was a Laurel and Hardy movie (probably *A Chump at Oxford*) that the boys watched being filmed, courtesy of producer Grinieff.

The d'Arbeloffs would spend the war years in the New York area, moving from one location to another as required by Vlad's and Sasha's work. Jobs with a company manufacturing ship engines brought the clan to Haddonfield, New Jersey, so Alex was enrolled at Haddonfield High School for a year. Then it was off to Mamaroneck, New York, and, for Alex, Mamaroneck's Bellows High School, where he graduated in 1945, editing the School's Yearbook and competing in track and basketball for the "Black and Blue" of Bellows High.

Just before he completed high school, however, his parents presented him and his brother Dee with a third brother, George — the first of the d'Arbeloffs to be born in the United States. The 17-year age gap between Alex and George was unusual enough, but stranger still was the fact that in the same year Sasha's wife Blanche also gave birth to a son, also 17 years younger than their oldest child! Thus the brothers Vladimir and Sasha d'Arbeloff each had three children of about the same ages, with the oldest and the youngest separated by 17 years. In each family, the two older children were born in Paris, the youngest in the United States.

The d'Arbeloff tale is remarkable for many reasons, not the least of which is the strong sense of family that was always present. Home to the d'Arbeloffs was wherever the family gathered, and the most remarkable characteristic of the saga is the constant regrouping of the family in one new setting after another. Parents and children moved from country to country as a unit, bonded together by Russian tradition and by the abiding friendship of the two Russian brothers, Vlad and Sasha. Vlad's oldest sons, Alex and Dee, continued the tradition. Their mother, Catherine Tiepolt d'Arbeloff, spent her twilight years at the Cambridge home of her son, Dee. Their father, Vladimir, died relatively young, in 1962, but Uncle Sasha lived to see his 101st birthday, and aunt Blanche, at 96, lives — and paints — in London as this is written.

As Alex was about to graduate from high school, he applied, at his father's insistence, for admission to only one college: MIT. Although his high school grades were only fair, he scored well in the New York Regents exams and was admitted to the Institute on July 1, 1945, as a member of the "Class of 2-48," which meant that he was scheduled to graduate in February, 1948. Like Nick DeWolf, he was subject to "acceleration" and moved directly from high school to college in July.

At MIT, Alex lived for his first two years at the Sigma Chi house on Beacon Street, then moved into a rented apartment on the same street, rooming with two classmates. He was a member of Tech's Debating Society and Team and also ran in intramural track meets. He was an average student, majoring in Business and Engineering Administration (known as Course XV), but, as in high school, he was able to bear down when necessary.

His accelerated schedule was derailed by a summer spent at ROTC camp (bunking next to Nick DeWolf) and by his decision to work for a summer at the Yale & Towne Company in Stamford, Connecticut, as a time keeper. He thus did not receive his degree until January, 1949. Upon graduation he accepted a job in Yale & Towne's Methods Department in Stamford, Connecticut. His salary was $40 a week.

MIT's ROTC program had given him a Reserve Commission, and in May of 1949 the nation decided that its security demanded the services of 2d Lt. d'Arbeloff. He reported for duty at the Army's ammunition depot in Joliet, Illinois. First, however, there would be a contretemps at Yale & Towne. The Company, seeing that d'Arbeloff was packing his bags, asked him if he would mind being laid off in order to help the Company effect a planning downsizing. Alex would indeed mind, and he insisted (against the strong advice of his superior) on taking his case to the Director of Industrial Relations, where he prevailed. The outcome was inconsequential, but the episode was noteworthy as a harbinger of confrontations to come.

Alex d'Arbeloff was not destined to write a glorious chapter in his country's military history. In fact, after four months at Joliet, his commanding officer, a General Holmes, delivered the following broadside: "D'Arbeloff, in the short time you've been here you've managed to antagonize every officer on this post." Holmes then found a colonel at a post in Ravenna, Ohio who was willing to take the troublemaker off his hands, and that is where the young lion served out his tour of duty. On a Friday in 1950 he received orders to report to a separation center. Two days later the Korean War broke out, and the following week d'Arbeloff was a civilian.

He returned to Yale & Towne, but the relationship that had soured on his departure never fully healed, and he left the following year to join watchmaker Benrus, as Assistant to the Manufacturing Manager, at a plant in Waterbury, Connecticut. While working at Benrus he boarded nearby at the home of the headmistress of Westover School, a formidable lady whose brother had been the consul in Asuncion during the d'Arbeloffs' stay in Paraguay.

D'Arbeloff spent almost four years at Benrus — a remarkable length, considering the fact that he was in a more or less steady state of rebellion. Yet Benrus made its mark on him,[1] principally

[1]It is less clear that d'Arbeloff made his mark on Benrus. In 1955, when a manager informed Benrus patriarch Ben Lazarus (for whom the Company was named) that "Alex d'Arbeloff was leaving," the estimable "B.L." answered, "Alex who?"

by exposing him to the work of the quality gurus of the day and to the tools of the quality trade, such as statistical quality control.

The departure from Benrus was followed by another improbable chapter in Alex d'Arbeloff's improbable journey. This episode would bring him to Brazil, at the invitation of Uncle Sasha, whose latest venture involved the construction of cement block houses in the fast-growing metropolis of Sao Paulo. Alex's mission was to manage the operation for Sasha, working with a local entrepreneur. For a year and a half, Alex lived happily in Sao Paulo — so happily, in fact, that he began thinking seriously of making Brazil his home. He added Portuguese to his language portfolio and was ecstatic when, while flying on a Brazilian flight, the local next to him asked him what part of Brazil he hailed from.

And why not Brazil? After living in Paris, Paraguay, New York, California, and Boston, Alex had learned that the keys to the kingdom were the same the world over: talent and a strong work ethic. Sao Paulo was vibrant and obviously growing rapidly. The climate was agreeable, the food good, the girls beautiful. Why, indeed, not?

The "why not?" was answered when he was called home to be with his father, who had suffered a serious heart attack. Vladimir recovered, but the family ties reasserted themselves during Alex's stay, and thoughts of returning to Brazil were put aside. Dee had graduated from Harvard, spent two years in the Navy, and earned an MBA from the Harvard Business School. George was nine years old. The family was now living in a comfortable home in Greenwich, Connecticut. Alex was still not sure what he would do with his life, but he decided that, whatever it was, he would do it in the United States, probably in the northeast.

From 1956 through 1960 Alex worked for three companies. The first was NRC Equipment, a maker of high vacuum equipment in Newton, Massachusetts. There he served as Production Control Manager, and when the Production Manager left the Company, Alex bid for his job. When he was refused, he asked for a position in sales. NRC told him he had no future in that area.

The next stop was Laboratory for Electronics, known as LFE, also in the Boston area. The job came through friend and fellow squash player Peter Redfield (later CEO at ITEL). At LFE Alex was Production Manager. He was fired after a year. ("They thought I was a pain.")

In 1958 he joined Artisan Metal Products, first as an assistant to the President (an MIT alumnus named Jim Donovan), later as manager of the firm's film-processing equipment division. The association with Artisan would end badly, the Company firing him in the Fall of 1959, but the experience was somewhat more productive than the other jobs, if for no other reason than that it sparked a long-lasting friendship with Nick Cedrone, a talented mechanical engineer who would contribute much to Teradyne's early success, both as a consultant and as the founder of a company (Daymarc) whose component handlers would be paired with Teradyne testers in test installations the world over.

Cedrone recalls the day he first noticed d'Arbeloff. "He was in the next office, talking to a customer who was all upset about something. And Alex was supposed to calm him down and make things all right. I didn't know Alex at the time, but I could hear him dealing with this angry customer, and he was masterful. So I went to the boss later and asked if this fellow d'Arbeloff could be transferred into the film processing division. His people skills were just outstanding. And they always have been."

How can the same person be judged a "pain" in one job and a master at dealing with people in the next? The most plausible answer lies in d'Arbeloff's difficulty in accepting a subordinate role. To his superior officers in the Army and to his managers at Benrus, LFE, and elsewhere, he was a chronic malcontent, constantly chafing at instructions he thought wrong-headed. But in situations where he was in control, there was no more skillful practitioner of human relations on the face of the earth. Undoubtedly, he felt himself in control while mollifying the Artisan customer.

At the moment, however, the world was not offering him a position in which he could be in continuous control, and frustration was building. It was particularly acute when, in the summer of 1959, he attended his MIT class's tenth reunion on Cape Cod. There he renewed old friendships with classmates who for the most part were well launched in promising careers and who had wives and children. Taking stock of his own situation — still single, with a checkered work background — he did not appreciate what he saw, and he fled the reunion, depressed, on the second day of the three-day outing.

As a young bachelor living in the Back Bay, d'Arbeloff socialized regularly and was a member in good standing of the Sitzenjibbers and regularly attended the group's parties. Here the celebrity status of fellow Sitzenjibber Nick DeWolf was the subject of much cocktail chatter, and d'Arbeloff was intrigued. Never close friends, Nick and Alex nonetheless had their shared ROTC experiences, and now each was part of the same social scene. There was more than enough common ground, in short, to warrant asking DeWolf for an introduction to his associates at Transitron. Of course, replied DeWolf, he would be happy to arrange an interview with Transitron's President, David Bakalar.

D'Arbeloff arrived for the scheduled appointment full of hope. This could be the turning point, the job that finally put him on the road to greatness.

David Bakalar, alas, never showed up for the interview. The moment came and went.

Shortly thereafter, DeWolf decided to leave Transitron to start his own company. Once again, the Sitzenjibber network alerted d'Arbeloff, who by then had forged another link with DeWolf: He was dating a girl who had been Maggie DeWolf's roommate in Boston. Again d'Arbeloff called DeWolf, and the two began meeting to discuss the broad outlines of the company that might be built around DeWolf's ideas.

Others were attracted by the opportunity to join forces with DeWolf, and the brainstorming sessions involved, in addition to d'Arbeloff, two friends of DeWolf, Don Sohn and Hank Rosen. As the discussions intensified, it became clear that the overriding issue on the table was DeWolf's choice of a partner for his new venture. Rosen, staking his claim, proposed a company headed by DeWolf and himself, with d'Arbeloff working as a sales representative, on commission. This was hardly an offer to excite d'Arbeloff. He was, after all, being asked to commit to selling products that didn't yet exist, for a company that didn't yet exist. He was of course willing to take risks, but this was too much. He stalked off, convinced that his business relationship with DeWolf had ended before it began.

The next day, however, Maggie DeWolf telephoned d'Arbeloff, asking him to be at the DeWolfs' Beacon Hill town house at five o'clock. It was an uncharacteristic intercession on

Maggie's part, and Alex was persuaded to leave his Marlborough Street apartment and walk over to Pinckney Street one last time. When he arrived, he found Nick decisive: Alex was the partner he wanted in starting his new company. The pair's skills were complementary, the chemistry seemed right. Thus, on that fateful afternoon in May, 1960, the son of Captain Maurice DeWolf, U.S. Navy, and the grandson of Admiral Tiepolt, Russian Imperial Navy, shook hands in Boston and agreed to set sail together into the uncharted waters of semiconductor testing.

PART TWO

The Exhilarating Sixties

3

• • •

Laying the Keel

The new company would make test equipment. There were other possibilities in DeWolf's deck of 3 by 5 cards, but his acquired mastery of semiconductor test techniques could not be ignored. He demonstrably knew more about the art and science of diode testing than any man alive. It remained only to build a company with staying power using that knowledge as a base.

DeWolf and d'Arbeloff had enormous advantages as would-be company builders. They were young enough (32) to attack problems with great energy. Both had first-class minds, and Nick was a bona fide technical genius, the kind other engineers regarded with awe. Most important of all, each had worked about 10 years in the real world. The experiences accumulated by d'Arbeloff at places like Benrus and LFE and NRC often infuriated him at the time, but there was no doubt that they made their impression, if only by teaching him how not to run a company. DeWolf had worked for a large company, one of the very best, and he had been present virtually at the creation of a successful start-up. Both had seen enough of industry to have a reasonable idea of what had to be done at the outset.

You needed information, lots of it, and in the pre-Internet era you collected it by sending away for catalogs and attending trade shows. And so the two flew off to the Wescon exhibit in Los Angeles, where they stuffed their bags with specifications of potentially useful devices and began collecting names and phone numbers for their Rolodex.

The seed money wouldn't be a problem, First, they didn't need that much. The embryonic company could operate on the fifth floor of DeWolf's Pinckney Street town house. Second, the two could afford (barely) to put up $25,000 of their own money. Third, Nick's reputation as a winner was good enough to attract an additional $175,000, mostly from family and friends.

Then there was the matter of a name. Years later, when asked where the name Teradyne originated, d'Arbeloff would often say something like, "Well, first of all, it had to have a 'd' in it." Such feeble reasoning was no doubt intended to disguise the fact that the name had no meaning whatever. "Tera" is the metric prefix for a quadrillion, or 10 to the twelfth power. A dyne is a unit of force, and a teradyne works out to about 1000 tons of force — a fact of absolutely no significance. The fact is, Teradyne was a word coined by the two founders simply because it sounded "high-tech." Megadyne was considered and rejected when it was found to be already in use, and Gigadyne just didn't sound right. So Teradyne it was.

Teradyne was a reasonably good choice for a name in 1960. What the founders failed to foresee was that a profusion of similar names would eventually cause the name "Teradyne" to lose whatever cachet it might have had at the beginning. Had the company been named "DeWolf-d'Arbeloff" or "d'Arbeloff-DeWolf" (as in "Hewlett-Packard"), uniqueness would have been ensured. But then, other problems might have arisen. On balance, one must acknowledge that the name Teradyne has neither contributed to nor detracted from the success of the Company.

It is a truism that nothing matters more in the creation of a new company than the formulation of a sound, detailed business plan. The Teradyne experience validates the thesis. The two founders hammered out their plan as carefully as Michaelangelo hammered out David. The essential strength of this plan is so central to the Teradyne story that it is printed in its entirety in the Appendix. The opening paragraph is worth excerpting here:

> *General Objective:*
> *It is the objective of Teradyne to design, manufacture, and market electronic equipment for industrial use... Electronic equipment for industry requires a high order of reliability and dependability. The high cost of down-time for maintenance and repeated calibration is sufficient to warrant more care in the design of such equipment and the use therein of higher quality components than in the production of consumer goods. The penalties to the user of undetected improperly functioning equipment may be many times the original cost of the equipment.*

The founders went on to say that the super-reliable components they intended to use had been designed for use by the military but were shunned by the consumer and industrial markets because of their high cost, while the military contractors were not organized to do business in the commercial arena. But the high cost of these components was an illusion, they held, because the true cost of production equipment lay, not in the capital cost, but in the penalty paid for unreliable testing and in the "down-time" required for calibration and maintenance.

This concept, so instinctively true today, was alien in 1960. In the catalogs and advertisements of the test industry's leading companies — Hewlett-Packard, Tektronix, and General Radio — the word "down-time" is nowhere to be found, nor is the economic value of reliable testing mentioned. The equipment offered by these manufacturers was designed primarily for use in the electronics laboratory, not on the production line. In the laboratory, the specifications of interest were accuracy and precision. The objective was the measurement of a given parameter against a standard — a resistor was so many ohms, a capacitor so many microfarads, etc. To DeWolf, the objective was not measurement, but test — a diode was good or it was bad. The tester would of course determine the goodness or badness by making measurements, but this was incidental and at any rate should ideally be invisible to the operator.

The shift of focus from measurement to test was intensified by the nascent semiconductor revolution. The diodes and transistors made by companies like Transitron, once their fabrication was mastered, would roll off the production line at the rate of thousands per hour.[1] Such high volume demanded high-speed, reliable testing. Moreover, the cost of good parts rejected or bad parts passed, or the loss of a tester for an hour or more because of malfunction could be disastrous in terms of lost revenue and damaged reputation, representing, in the words of Teradyne's business plan, "many times the original cost of the equipment."

[1]Vacuum tubes were also produced in high volume on automated production lines, but semiconductor mass fabrication techniques would redefine the meaning of "high volume."

Surely, the reader will say, instrument makers as successful as Hewlett-Packard, General Radio, and Tektronix must have foreseen the impact that semiconductors would have on their businesses. But the semiconductor, although functionally analogous to the vacuum tube, represented a major discontinuity in the electronics industry, a discontinuity for which the engineers and managers of the established companies were poorly prepared. (It is worth inserting here the telling fact that not one of the leading makers of vacuum tubes of the 1950s appears on a list of today's leading semiconductor makers.) It is significant that neither Teradyne founder had worked for one of the "big three" test equipment companies of the day. In fact, Teradyne was eight years old before it hired its first employee from any of these companies. The key engineers who created the first waves of products were all, like DeWolf, from the semiconductor industry.

Louis Lemle, the New Orleans lawyer who was DeWolf's father-in-law, was helpful in the drafting of the business plan, and he was especially valuable when it came time to work out the arrangement between the two partners. Lemle also was able to use his legal ties to find the Boston lawyer the two would need to launch their new company. He was Eugene Lyne, of Boston's Lyne, Woodworth, and Evarts, and he was to play a major role in Teradyne's first decade, initially as outside counsel, later as employee and officer.

If one day must be chosen as Teradyne's birth date, it is September 5, 1960. The place was the fifth floor of the DeWolf's Pinckney Street house. The next few months saw a whirlwind of activity. The Company was incorporated in Massachusetts on September 23, issuing 2000 shares of $10 par value stock for a total cash consideration of $20,000. Of these shares, Nick DeWolf held 1500, representing his investment of $15,000, while d'Arbeloff held 500, for which he paid $5000, and had another 500 reserved for future purchase at the same price. Thus the ownership of the new company was initially split 60-40 between the two. Almost immediately thereafter, family and friends were offered the chance to invest $150,000, in the form of 1500 shares of $100 par value cumulative convertible preferred stock. Virtually all the invitees accepted, and among these earliest investors was DeWolf's uncle, Henry Watts, a successful New York broker, former Chairman of the Board of Governors of the New York Stock Exchange, and, in later years, a member of Teradyne's Board.

Also participating in the first round of financing, with an investment of $10,000 for 100 preferred shares, plus an option to purchase 1800 common shares, was American Research and Development Corporation, the first publicly owned venture capital company. ARD, headed by the legendary General Georges Doriot, had also invested in a small Massachusetts-based computer company called Digital Equipment Corporation. In 1960 Teradyne was one of 17 new ARD investments. Although ARD earned wide fame for its ability to pick winners, few of its investments paid off as handsomely as did Digital and Teradyne. Of course, the whole venture capital game was based on the hope that one investment in ten would hit the jackpot.

With the first rounds of financing behind them, and with Pinckney Street growing less suitable as the site of a budding electronics manufacturer, DeWolf and d'Arbeloff began looking for a home for their new company. At the time, the most popular venue for high-tech start-ups was Route 128, the circumferential highway around greater Boston. Moreover, many of Teradyne's likely customers were located on or near the highway — companies like Sylvania, Clevite, Transitron, and Raytheon.

But DeWolf liked living on Beacon Hill, had no intention of moving, and wanted to be able to walk to and from work. Therefore Teradyne's first home would be in the City. D'Arbeloff, who lived in the Back Bay, readily agreed. Rentable space was found on Summer Street, in the heart of Boston's downtown shopping district. It was an improbable site for an electronics company.

In years to come, as Teradyne expanded and brought thousands of good jobs to the downtown area, urban planners and social activists often praised the sense of civic duty that they assumed prompted Teradyne to place its headquarters within easy commuting range of the inner city's residents. While the results were indeed impressive (Teradyne's minority employment was generally much higher than that of other Massachusetts high tech companies), the founders always turned aside such praise, insisting that walking to work, not doing good, was the whole idea. Although that is certainly true, there is a message here for urban development activists: Instead of trying to attract companies, attract the kind of people who start companies.

In November 1960, Nick and Alex opened for business on the third floor of 87 Summer Street, above Joe & Nemo's hot dog stand. The same month, a newly elected president, John Kennedy, spoke of the "passing of the torch" from one generation to the next. Although it wasn't clear at the time, the torch was also being passed from the old electronics industry — led by companies like RCA, GE, and Sylvania — to the new, represented by the likes of Fairchild Semiconductor, Texas Instruments, and Transitron. Years later, the torch would again be passed from Fairchild to Intel, National Semiconductor, and Advanced Micro Devices, while Transitron talent would desert that Company to form dozens of new companies and to add measurably to the technical prowess of Teradyne.

On the surface, the country was coasting along on a smooth course in 1960. Korea was behind it, and the agonies of Vietnam had not begun. Eisenhower's legacy was a solid economy, and most Americans were satisfied with their lot. They watched "Gunsmoke" and good movies like "The Apartment." The Celtics won the NBA playoffs (again), Miss America was Miss Mississippi (again), and the Red Sox (again) did not win the pennant. The only disturbing note was the deepening of the Cold War, but even this did not seem to cause great anguish among the general public.

Nor did it particularly upset DeWolf and d'Arbeloff, who focused almost maniacally on their new Company, pushing aside all distractions. For d'Arbeloff, a bachelor, this was no great sacrifice, and he threw himself into the details of company-building with total dedication. DeWolf, blessed with a heroically tolerant wife, was also able to indulge his all-consuming interest in semiconductor testing.

So they settled into 87 Summer Street. The first employee, a secretary named Ellie Hight, was hired, partitions were built, and Alex knuckled down to the task of learning what made semiconductors tick. For, although he had graduated from one of the world's great engineering colleges, at the time of his graduation the transistor had barely been invented, and his working knowledge of semiconductors was nil. But he was a quick learner, and after a few months under DeWolf's tutelage he would be writing long letters to prospective customers, advising them of the proper techniques for measuring the leakage current of diodes.

4

• • •

The Diode Tester

Teradyne's first product, naturally enough, would be a diode tester that would embody many of the principles promulgated by DeWolf at GE and refined by him at Transitron. The new tester would be reliable, easy to use, and give semiconductor makers the throughput they needed to compete in what was even then turning into a brutally competitive business.

(DeWolf was well qualified to launch his new company with a transistor tester, but he thought the diode a surer bet. When exploiting a new technology, he felt, it was wiser to enter at the simplest level and work up.)

Hearing of the new Company's intentions, and aware of DeWolf's reputation as Transitron's diode-meister, Raytheon gave the two entrepreneurs the opening they needed. Raytheon was one of the pioneers in the semiconductor industry, producing transistors for hearing aids and diodes for a variety of consumer applications. On January 4, 1961, it invited Teradyne to bid on a "Semi-Automatic Diode Test Set," which would perform five electrical tests on each diode, executing the tests fast enough to deliver a throughput of at least 1200 diodes per hour.

At the time, Nick DeWolf was clearly better qualified than any engineer in the world to respond to such a "request for quotation," and within a week he and his partner were exchanging visits with Raytheon's Purchasing Agent, in Boston and at Raytheon's Semiconductor Division in nearby Newton.

Nick DeWolf was clearly not interested in designing a tester that would merely meet Raytheon's specifications, which, by his standards, were misguided anyway. "Teradyne arrogance" was thus present at the creation, because it was in DeWolf's genes. It manifested itself in the beginning and for years thereafter in the abiding belief that Teradyne knew more about its customers' testing problems than the customers did. At least in those early years, that was almost always true.

The tester that would take shape over the next six months would embody all the principles of automatic test design that DeWolf held dear since his GE days. It would be easy to operate, reliable, fast. DeWolf designing the Company's first product was a quarterback throwing a 60-yard touchdown pass on the first play of the game.

Raytheon had specified a throughput of "1200 diodes an hour." DeWolf scratched out "1200" and substituted "3600." To speed up testing, Raytheon had specified that the tester stop as soon as any test failed. DeWolf promised to run *all* the tests (thus generating useful data on rejects) without any productivity penalty. Raytheon had specified a test time of 2.5 seconds from the time the operator pushed the "start" button until the "accept" or "reject" signal was displayed. DeWolf crossed out "2.5 seconds" and penned in ".75 sec." Raytheon had stipulated a one-year warranty. By the time the first diode tester was shipped, DeWolf was guaranteeing the product for 10 years.

In April, DeWolf hired his first engineer, Jozef Hendriks. Hendriks, a native of Holland, had designed in-house testers at Transitron for DeWolf, and the two shared the same design philosophy, the same in-depth knowledge of what was going on, not only in each circuit, but in each circuit element.

DeWolf and Hendriks knew that one of their toughest challenges was the measurement of leakage current, a notoriously unstable parameter that brought other testers to grief. A tester operator measuring leakage current might wait an interminable length of time while a meter crept slowly upscale to some horribly leaky value. It was not uncommon for the tester meter to be slower than the diode itself, making tester productivity hostage to the ballistics of the meter movement. In some cases, the meter needle would overshoot and dither around for a few milliseconds before coming to rest — another productivity robber.

The solution, DeWolf decided, was to have a meter manufacturer custom-design a movement to his specifications, which would include a damping circuit consisting of two coils, one to carry current, one for feedback. After repeated discussions with meter manufacturers, d'Arbeloff cut Teradyne's first purchase order, for the meter movement that would, with luck, bring a new order of speed to diode testing. Feeling awkward about generating "purchase order No. 1," he instead marked it "purchase order 141."

Once the meter movement arrived, DeWolf attacked it with characteristic intensity, applying current, timing the pointer's travel with a stopwatch, tweaking the circuit, clocking it again, and little by little losing faith in any meter's ability to operate fast enough for his diode tester. Early in the game, therefore, he made a fundamental course correction: Forsaking arrogance, he accepted Raytheon's suggestion that the tester be made fully automatic, indicating, for each test, whether the measured value was within specification. The "diode tester" thus became the "go/no-go diode tester" because at a crucial moment DeWolf listened to his customer. The operational amplifiers DeWolf had used to drive the meters were replaced by comparators, which activated green (pass) or red (fail) lights. A string of green lights meant the diode had passed.

The business end of the tester was the clip into which the operator would feed each diode, and DeWolf's experience taught him that the best-laid plans of any circuit designer can be derailed by something as simple as a bent connector or a faulty test fixture.

A diode has two terminals, or leads, and designers of other diode testers therefore equipped diode clips with two terminals, one for each lead. The two-terminal clip, unfortunately, had a serious flaw: The contacts' resistance is comparable to that of the diode itself and, fluctuating wildly, cannot be calibrated out of the measured results. DeWolf's solution: a four-terminal, so-called "Kelvin" connection, in which current is pushed through one pair of terminals while voltage is measured at the other pair (or vice versa). DeWolf was of course familiar with the work of Lord Kelvin, who had fretted about the series resistance of telegraph cables a century before. Hendriks knew all about Kelvin, too, as did any truly competent electronics engineer. Yet Teradyne's four-terminal diode clip was a unique feature of its testers for a surprisingly long time.

There still remained the task of translating DeWolf's Kelvin circuit into a physical fixture that would capture each diode, connect it to the four terminals, and allow it to be released without damage. The conventional fixture, which was called a "J clip," required the operator to hold the two ends of the diode in two hands, insert it into the clip, push the "start test" button, and remove the diode, again using two hands. Often the operator would pull one end before the other, and the diode lead would be bent. Firm in his view that, for want of a good clip, the whole tester would be lost, DeWolf looked for mechanical help.

He did not have to look far. Nick Cedrone, d'Arbeloff's colleague at Artisan Metal Products, had left Artisan in 1959 to found his own company, called Daymarc, to manufacture temperature controllers. In mid-1960, as Teradyne was being conceived by DeWolf, d'Arbeloff brought Cedrone to DeWolf's Beacon Hill apartment. DeWolf explained the shortcomings of available diode clips. "There must be a better way," he prompted Cedrone. D'Arbeloff contributed an idea for a vee-shaped fixture that would provide the needed four-terminal connection, and Cedrone's wheels started turning.

Soon Cedrone was back with a prototype, which greatly improved on and elegantly implemented the vee-shaped fixture. DeWolf was ecstatic and, after a number of refinements, the Cedrone diode clip would sit smartly on the front deck of Teradyne's first product. Even before the first diode tester would be shipped, hundreds of diode clips, "made by Daymarc, offered by Teradyne," would be sold under the terms of an agreement between the two companies signed in May, 1961. The clips, which were much more advanced than anything on the market, were sold to builders of in-house testers who were attracted by the four-terminal design and other features of the clip. Among these features was a magnet that literally pulled each diode from the operator's hand, so that it was necessary only to drop the part somewhere in the clip's vicinity, and the magnet would do the rest. With Teradyne's diode tester, moreover, the operator didn't even have to worry about dropping the clip into the clip backward. As each diode was tested, a latching circuit in the tester remembered its polarity. If the next diode it saw wasn't oriented the same way, the tester automatically reversed its own polarity.

The diode clip established Teradyne's reputation for innovation even before its first tester was shipped. It also cemented Cedrone's relation with Teradyne.

DeWolf felt strongly that his testers should be an order of magnitude more reliable than anything else on the market, and he wanted the mechanical design of his products to reflect that quality. An allusion to Mack Trucks was common, and one of the Company's early ads (scotched before it ran) built on the simile, with a twist, boasting that "Mack is the Teradyne of the automotive business."

In the quest for mechanical excellence, DeWolf saw in Cedrone a soul mate, and the latter was hired as a consultant to design the cabinet for the first tester. He also designed the sealed modules that would house the tester's innards — sealed to prevent tampering by customers, who, arrogant Teradyne insisted, would only get into trouble if you gave them access to the circuit boards. The Cedrone-designed modules were, as DeWolf ordered, castings that you could drive a truck over, even a Mack truck.

While Cedrone was designing the module castings, Joe Hendriks was designing circuits that would reliably and accurately measure the leakage currents and forward and reverse voltages of temperamental semiconductors — using the very same temperamental semiconductors in his circuits. The voltages were often painful to generate and manipulate rapidly, and, per one of DeWolf's first laws, could not under any conditions be allowed to destroy the device under test. Hendriks carefully wound his toroids and designed oscillators to produce the safe, gently varying voltages and currents needed, components were soldered onto circuit boards by d'Arbeloff (newly handy with a soldering iron), and the boards were carefully nestled inside the sealed modules.

The front panel of the diode tester would be the locus of one of DeWolf's most ingenious innovations. The panel of the traditional test instrument was a heavy metal plate with a gray- or black-crackle painted finish and with the necessary legends engraved on the surface. This classic look was handsome to look at but costly and time-consuming to produce and utterly inflexible. The finish, moreover, was vulnerable to scratches.

DeWolf of course cared nothing for the traditions of test instrumentation, and he set out to produce a front panel that would look good and be (for him) easy to produce and easy to change. His long-time passion for photography led him to consider applying the necessary lettering to a sheet of Mylar film and laminating the Mylar to an aluminum front panel. Thus, if one desired to change the word "milliamperes" (DeWolf preferred spelling out the names of units) to "microamperes," one simply produced a new Mylar sheet with the new lettering, executed with Letra-Set stick-ons.[1] Using this technique, Teradyne could produce a new panel from scratch in a half hour.

This was more than a cute touch; coupled with the tester's modular construction, it was a diabolically clever way to produce customized testers that appeared to be standard catalog models. Within a year of its founding, Teradyne was offering its customers a range of products that seemed to rival in scope

[1]The Mylar laminating process was much trickier to produce than it sounds here. The "magic" heat-sensitive glue/paint, if not just right, would result in delamination, and of course appearance demanded that there be no bubbles or ripples. But to DeWolf, who had dry-mounted 16 by 20-inch photographic prints to Bristol board by the hundreds, it was familiar territory.

the entire Hewlett-Packard catalog. In fact, the blizzard of type numbers that one finds mentioned in the early Teradyne literature simply represented reconfigurations of a handful of basic modules, each variation individualized by a new sheet of Mylar bearing new front panel markings. What's more, the rugged Cedrone-designed cabinet and the scratch-resistant Mylar presented a clean, modern appearance, which stood in contrast to the fusty look of other equipment on the market.

The Company's ability to turn out new front panels on the spot helped it to respond to customers' needs, or invitations to bid, with amazing speed. Many years later, much would be written about the importance of "time to market" in the fast-changing equipment business, and one day Michael Dell would build a major computer company around the "build to order" theme. But no test equipment company ever compressed time-to-market as effectively as did early Teradyne.

DeWolf wanted Teradyne's testers to require a minimum of down-time for calibration and repair, and this in turn led to an entire code of engineering conduct, including a long list of "Thou Shalt Nots":

> *Thou shalt not use potentiometers, because they will inevitably be misadjusted. Design the circuit right in the first place, and you will not need such crutches.*
>
> *Thou shalt not use trimmer capacitors, for the same reason.*
>
> *Thou shalt not use forced-air cooling. Fans need filters, which need periodic replacement. The user will forget to replace the filters, and the tester will overheat and destroy itself.*

Thou shalt not allow the tester to damage components under test, even if the user misprograms all the controls.

Thou shalt not test at frequencies beyond those necessary to characterize the device, because speed kills.

"Speed Kills" became a DeWolf mantra, and it was the title of a trade magazine piece written by him years later.[1] The phrase owes its origin to the drug wars that were raging at the time at, among other places, DeWolf's Beacon Hill. The point was, that whether one is dealing with amphetamines or cars or motorboats or semiconductors, speed is usually a bad idea. In the semiconductor context, speedy devices tend to run hot, which causes a multitude of problems. Of course, the quest for ever-higher speeds would become irresistible, because the devices to be tested ran ever faster, and Teradyne would in time take up the chase. But during DeWolf's reign, the testers would be designed with relatively slow components — but with blazing fast throughput, thanks to DeWolf's (and Hendriks's) ability to extract a maximum of useful measurement data from a minimum number of operations. DeWolf may have been preaching "speed kills," but the customers knew that, where it counted most — number of devices tested per hour — Teradyne was the fastest gun in town.

The new diode tester would showcase all of DeWolf's design principles, but from the outset these principles were not about a diode tester, but about a company. DeWolf and d'Arbeloff were already thinking ahead, to the transistor tester that would inevitably follow, as well as testers for Zener diodes and resistors. There seemed almost no limit to the number of

[1]"Nick DeWolf of Teradyne Speaks Out: Speed Kills," *EEE*, August 1970

testers one could produce if one had the right test modules, all-purpose cabinetry, and the ability to create new front panels quickly out of whole cloth — or whole Mylar.

One also needed the right components, and DeWolf knew that his tester could be only as reliable as its least reliable component. The early months of 1961 saw d'Arbeloff, under DeWolf's direction, writing to hundreds of component makers, asking for catalogs, specification sheets, samples, and quotes. If the perfect component did not exist, DeWolf would not hesitate to ask a supplier to redesign something to meet his requirements. Switches, reed relays, cable, printed circuit material, pilot lights, dials, capacitors, resistors, toroids — they were all important, and DeWolf spent countless hours studying their characteristics in order to ensure that what Teradyne purchased would not compromise his standards of reliability. This was especially true of reed relays, upon which the operation of automatic testers heavily depended.

Most of the components listed in DeWolf's short list of approved parts were distinctly non-exotic. His favorites were parts that were so thoroughly mainstream that they could be bought at the local Radio Shack (whose first store was in downtown Boston, not far from Teradyne). The parts list thus included mostly garden-variety electrical components, plus a carefully chosen assortment of reed relays and input and output transistors. The total list was short — fewer than 200 electrical parts — and DeWolf was determined to keep it short. This obsession added to the reliability of Teradyne's testers, and it also meant that the Company ran on a lean inventory mix — a habit that would serve it well throughout its history.

By March, Teradyne's letterhead was sporting a Company logo: a stylized letter "T" inside a turquoise triangle. The triangle and the stylized "T" are long gone, but the use of turquoise as the "company color" persisted for decades thereafter. Why turquoise? According to DeWolf, because the color is associated with the American Indian, and he had once been told by his parents that he was part Native American, in order to stimulate his sense of patriotism. This turned out to be utter nonsense, but as a story it stands up at least as well as does the explanation of the Company's name.

The new diode tester now taking shape was initially called the D5, then the D5A, then, finally, the D133. Thus began a numbering system that would serve the Company for decades. Each tester would be given an alphanumeric designator made up of a letter prefix standing for the unit to be tested (D for diode, T for transistor, Z for Zener diode, etc.) followed by a three-digit number. So far, routine enough. But the numbers themselves would be invested with a special quality: They would be prime (divisible only by themselves and 1) to suggest, at least to those in the know, the indivisibility of the products. Teradyne would sell entire testers, not test modules. The rule stuck, and most of the major testers of the DeWolf era would be christened with prime numbers.[1]

Teradyne's diode tester, like all the other testers in the back of DeWolf's mind, was intended to be used on the production line, where speed of operation (throughput) was paramount. The magnetic clip, the left- and right-hand start buttons, and the short test time combined to allow the tester to process diodes as fast as a fast operator could drop them onto the clip and then place the

[1]Years after DeWolf's departure, many non-primes would infiltrate the system. Then, in the 1990s, the engineers who created Teradyne's first VLSI test system would pay tribute to their Company's founder by calling the system the J941. 941 is not only prime; it is an EMIRP — a number that is prime backward and forward (i.e., 941 and 149 are both prime). Other "EMIRP" systems included the J953 and the J967.

good ones in the "accept" bin, the bad ones in "reject" bins corresponding to failed tests. Since the tester needed only a quarter of a second to execute all the tests on each diode, there was no way a human operator could keep up with the tester, and if the tester's inherent speed was to be fully exploited, the handling and the sorting would have to be just as automatic as the testing.

A few small companies produced handlers, and in May 1961 Teradyne signed an agreement with one of them, Transistor Automation Corporation (TAC).[1] Under this agreement TAC would develop handlers for use with Teradyne testers, and Teradyne would serve as exclusive sales agent for these handlers. Nothing was to come of this agreement or of discussions held later in the year with another handler manufacturer, Universal Instruments. But his association with Teradyne had convinced Nick Cedrone that device handling equipment presented a much greater opportunity than did temperature controllers, and Daymarc veered into the semiconductor seas, eventually to become a world leader in the semiconductor handler business.

On April 12, 1961, with Raytheon's final specifications in hand, d'Arbeloff quoted a price of $5175 for the diode tester. The initial quote, delivered in February, had been $5750, and the lower price reflected DeWolf's confidence in its design and in its manufacturing costs. Still, $5175 was far higher than the price conventional wisdom assigned to such testers, and the initial reaction of the market was along the lines of "You can't be serious." But DeWolf and d'Arbeloff felt strongly that their new, industrial-grade testers would quickly pay for themselves in added throughput and reliability. It was not an easy sell, but the new Company had two powerful weapons: DeWolf

[1]TAC would in time become Teledyne TAC, prompting some confusion between Teradyne and Teledyne. Some Teradyne wags would deal with this by explaining that "Teledyne is our Japanese subsidiary."

("Transitron's best salesman") and d'Arbeloff, who had discovered what his superiors at National Research had missed: He was a born salesman.

A two-page trip report, dated 14 June 1961, showcases d'Arbeloff's tenacity. Hearing about the four-terminal diode clip from Raytheon, CBS Electronics asked for a demonstration, and d'Arbeloff drove to the CBS plant in Lowell. After conducting a series of measurements to CBS's satisfaction, d'Arbeloff attached the clip to an in-house tester and waited while the operator tested 200 diodes without a single problem. "The demonstration went exceptionally well," wrote d'Arbeloff. "The production foreman liked the clip and everyone was suitably impressed." Then came more tests with the tester coupled to a Universal handler. Then a discussion with the Chief Industrial Engineer. Then, after some debate about price, the payoff:

"They decided to buy one (1) clip." (The clip was priced at $65.)

In August, Teradyne shipped its first tester, Raytheon's D133. The achievement was duly noted in a letter to stockholders — for the most part, relatives of Nick DeWolf.

The same month, the Company was quoting (to Microwave Associates) a Silicon Transistor Classifier for $3600 and a Silicon Transistor Parameter Tester for $4500. In September, a letter from DeWolf to Robert Hughes of Digital Equipment Corporation quoted an "Automatic Computer Transistor Test Instrument" at $4950 and an "Automatic Diode and Rectifier Test Instrument" for $4880, in both cases with quantity discounts. To Raytheon, also in September, went a quote for an "Automatic Computer Logic Module Test Instrument" for $3875. To Clevite, Transitron, and Sylvania went quotes for a

"Compensated Precision Automatic Diode Forward Test Instrument," at $4925. An "Automatic High Voltage Inverse Classifier" was offered to Sylvania for $4450. All this from an engineering staff that could be counted on the fingers of one hand, with fingers to spare. Clearly, the mixing and matching of test modules was in high gear. DeWolf had invented, not just a tester and a philosophy of testing, but a system for generating new products at a rate that had never been seen before in the test and measurement industry. Each tester quoted existed only in DeWolf's mind; as soon as the customer ordered it, it would spring to life in Cedrone's all-purpose cabinet, and a sheet of Mylar with the necessary legends would be bonded to the panel. To the customer, it would look for all the world like something Teradyne had been turning out for years.

And yet, as any engineer knows, it is devilishly hard to design simplicity into a product. In this case, each test module represented hundreds of hours of painstaking design work by DeWolf and Hendriks. The challenges mounted as the two broadened the scope of their efforts. The first diode tester, for instance, was designed to test germanium diodes. But silicon was fast becoming the preferred semiconductor material, and Teradyne therefore promptly developed a tester for silicon diodes, hurdling a technical barrier: While germanium diodes were characterized by high, easily measurable leakage currents, silicon diodes dealt in much lower currents — microamperes or even nanoamperes rather than milliamperes — where the measurements were more difficult.

With the diode testers launched, Teradyne's sights were now on other targets: transistors, Zener diodes, rectifiers, resistors — the whole universe of electronic components could be their oyster, if the Company played it right. Feeling sure they had a winning formula, DeWolf and d'Arbeloff began recruiting the engineers and technicians they would need, not only to design the new products, but to test them and (since the modules could not be opened by customers) to repair the small number that would fail. The initial recruiting effort focused on Transitron, where the talent was first-rate, equipped with a solid working knowledge of semiconductors, and well known to DeWolf. Transitron was also under increasing competitive pressure, and many of its engineers had begun circulating their resumes. Teradyne was not the only beneficiary of the Transitron exodus; Transitron engineers soon began showing up at a number of east- and west-coast semiconductor companies, and several started their own ventures.

Among the Transitron engineers recruited by Teradyne in 1962 was Bill Attridge, a hands-on circuit designer who would serve as a key member of the Company's technical staff for many years.

It is notable that despite the Bakalar brothers' understandable resentment of the way in which other companies were picking clean their Company's carcass, Nick DeWolf, almost alone among the predators, was able to stay in that Company's good graces, and Transitron would be among Teradyne's most loyal customers for years.

In November, Teradyne made its trade show debut at NEREM (Northeast Regional Engineering Meeting), a Boston convention sponsored by the Institute of Radio Engineers. Wishing to demonstrate their diode tester in a practical setting, DeWolf and d'Arbeloff arranged to share space in the booth of Universal Instruments, a maker of device handling equipment. Having an actual production machine doing real work in a trade show booth was a fairly novel idea at the time; most exhibits were static displays of instruments or components. The risk was that the handler might jam, turning the demonstration into a fiasco. And the Universal machine, known as the "jaw handler" because of its appearance, indeed did start jamming right from the start. But DeWolf was prepared, and, armed with a wire coat hanger, he extracted each jammed diode from the handler's jaws, littering the floor with tiny pieces of glass. This did not endear the Teradyne team to the Universal representative responsible for the booth, and the tension mounted further when DeWolf started eating a sandwich in between thrusts of the coat hanger. The agitated Universal representative reminded DeWolf that food was not allowed in the booth, and DeWolf responded with an ear-to-ear grin as he continued munching.

Jammed diodes aside, the trade show was a successful outing for the young Company. Although NEREM was essentially a local exhibit, the Boston area was at the time no less a center of semiconductor activity than was San Francisco. (The name "Silicon Valley" would not be coined for another 10 years.) Transitron, Sylvania, ITT, Sprague, CBS, Clevite, Raytheon, and others were all in the business, and all sent their people to NEREM to kick the tires.

The temporary alliance with Universal paid other dividends. That company had received an order from Corning for a handler to be used in the testing of resistors, and Teradyne was encouraged to design the tester. The sale was made, and a few Teradyne resistor testers wound up on Corning's production line, but the real payoff in resistor testing was to come later, when Teradyne would hit the jackpot at Allen-Bradley.

5

• • •

A Product Line Takes Shape

As 1961 ended, the two founders had to be pleased with their first full year in business. The offering circular they had prepared for potential investors in 1960 included a business plan that showed first-year revenues of $50,000. Actual revenues for the year were $47,000. The Company was in the red, of course, but working capital was not an issue; the Company had still not drawn on a $100,000 line of credit from American Research & Development.

The ARD connection was valuable to Teradyne as a door-opener and as a sign of impeccable credentials. The investment community was well aware that another ARD company, Digital Equipment Corporation, was a comer, and in some quarters it was whispered that Teradyne might be "the next DEC." ARD's president, General Georges Doriot, became a mentor, and he and d'Arbeloff were occasionally found chatting in French at a Bastille Day gathering of Boston's French community. Early in 1962 Teradyne added an ARD Vice President, Bill Congleton, to its Board.

The year also marked the end of d'Arbeloff's bachelor status. In April, at a Sitzenjibbers' party, he met Brit Jepson, the daughter of Sunbeam Corporation's Vice President for Engineering and the small-appliance industry's most prolific innovator, with some 500 patents to his credit. Brit was an engineer in her own right, having graduated first in her class at Stanford, where she was, moreover, the first woman ever to be awarded a degree in Mechanical Engineering. Then she earned a Master's Degree at MIT. She was stunningly beautiful as well as brainy, and d'Arbeloff wasted no time in pressing his suit. The two were married on September 9, 1962.

In the quarter that followed — the fourth quarter of 1962 — Teradyne crept into the black for the first time. Just before the year ended, the Company shipped its first resistor tester to Allen-Bradley. And the diode testers, Teradyne's bread and butter, were the talk of the industry.

Having achieved profitability, the Company decided that it liked the taste, and throughout 1963 d'Arbeloff increased prices as necessary to stay comfortably in the black. (For the year, Teradyne would net $46,000 on sales of $431,000.) Resistance to these increases was minimal, as customers began to see the economic advantages of Teradyne's new breed of testers. Word spread quickly about the new Boston company whose products never broke down and were covered by an unheard of 10-year warranty. More engineers were hired, the line was expanded to include transistor testers and Zener diode testers, and Allen-Bradley was marveling at the resistor tester it had just received from Boston. Allen-Bradley was known world-wide for the quality of its products, and there was an instant cultural bond between the resistor maker based in downtown Milwaukee and the tester maker from downtown Boston. DeWolf had moreover studied the stripes off the customer's products, and Teradyne's resistor tester was uniquely able to test A-B's nonlinear resistors at just the right voltage — a difficult but very important proposition.

In the course of the year, Teradyne issued quotes to prospective customers for the following products: the D106, D118. D119, D121, D122. D125, D141, D142, D149, D152, D157, D158, D161, D166, D169, D171, D197, D199, and D211 diode testers; the R103, R104, R105, R106, R126, R127, R129, R131, R133, R134, R138, R142, and R144 resistor testers; the T115, T135, T138, T139, T140, T154, T155, T161, T162, T163, T172, T175, and T202 transistor testers; the Z111, Z131, Z137, Z138, Z180, Z190, Z191, Z193, and Z194 Zener diode testers, and a new line of scanners, the M104, M107, and M109. Granted, most of these represented only minor variations of a few basic themes; it was nonetheless an astonishing output from an engineering team that consisted essentially of DeWolf, Hendriks, and Attridge. The manufacturing staff consisted of a supervisor named Ed Reddy and a small crew of assemblers.

DeWolf, totally absorbed in engineering, was happy to leave everything else to his partner, and d'Arbeloff was up to the task, busying himself with the details of leasing space, arranging bank loans, hiring help, and selling the products, either through his own efforts or by directing an expanding network of representatives. At the time, fortuitously, Hewlett-Packard decided to build its own direct sales force, and a number of excellent H-P representatives were cast adrift. Some of these were signed up by d'Arbeloff. As 1963 began, the Teradyne sales network included Applied Measurements (Massachusetts), SONCO (Pennsylvania), Gemini (Long Island), Bard Associates (Midwest), and Kittleson (California). Later in the year, at the NEC (National Electronics Conference) trade show in Chicago, d'Arbeloff interviewed three prospects to take over the Midwest territory, finally settling on Cozzens & Cudahy. The association would blossom in the following years and would lead Teradyne down new paths, as we will relate in due course.

The task of selling testers and coordinating a growing sales network was daunting even for d'Arbeloff, who was then generating a sea of correspondence, using the same modular approach that worked so well with the testers ("diode paragraph #5, "standard diode clip paragraph #3," he would instruct the typist). An assistant sales manager was hired to oversee the east coast. He was Dick Lucey, an ex-Transitron salesman working for a Transitron spin-off. Once again, Teradyne had recruited a "semiconductor type" rather than a "test equipment type," a pattern that would be repeated often, usually with great success.

Lucey's earliest challenges involved the defense of Teradyne's unique approach to system documentation and support. Unwilling to make customers privy to the secrets of his test modules, DeWolf constructed an ingenious rationale to keep them at bay. Its essence: You think you need schematic diagrams to service your testers, but you don't. What you really want is not a diagram, but a working tester. So if your tester breaks down we will ship you a replacement module faster than you could fix it yourself. Better yet, buy a few spare modules, which you can plug in when you need them.

The entire service program was centered on the modularity of Teradyne's testers and on the fact that one could replace any module with a spare without having to make any adjustments whatsoever. This feature was unique in the test instrument business and remained a Teradyne exclusive for many years.

Schematic diagrams did exist, of course, but these were subject to another DeWolf dictum: Every diagram had to be drawn on an 8-1/2 by 11 inch piece of paper, suitable for reproduction on an office copier. The diagrams were also to be drawn free-hand by the engineers, rather than by draftsmen.

Backed by a 10-year warranty, Teradyne's approach to service and documentation sounded persuasive, and in fact Teradyne's growing sales proved that many customers were persuaded. But the larger, more bureaucratic companies, especially those accustomed to military documentation, would balk and demand conventional schematic diagrams. To these demands, the farthest DeWolf was willing to go was to promise to place schematic diagrams in escrow with an attorney (Teradyne's) against the unlikely possibility that Teradyne might some day cease to exist. Lucey remembers spending untold hours defending "the Teradyne way" with customers, and certainly d'Arbeloff spent even more energy on the subject.

As 1963 came to a close, Teradyne was solidly profitable and growing fast enough to attract a number of inquiries from would-be investors. If there was a weakness, it was the fact that Teradyne's was essentially a custom business, made to appear non-custom through the modular architecture and front-panel wizardry we have already described. What it lacked was a high-volume product to generate the juicy incremental profits all manufacturers dream about. That breakthrough business fell into the Company's lap, in the form of an order for 85 resistor testers, all identical, from the world's leading resistor maker, Allen-Bradley.

D'Arbeloff had sold a few resistor testers to Allen-Bradley in late 1962, and the high throughput and reliability of these testers impressed the engineers at the Milwaukee-based company.

Quotes for quantities of testers were solicited and specifications were fine-tuned. The negotiations were carried on by d'Arbeloff and DeWolf with the aid of the Company's new midwest representative, Cozzens & Cudahy. Finally, the deal was closed, and Teradyne had an order for 85 testers, making the resistor tester Teradyne's first big "runner" and virtually guaranteeing the Company a profitable 1964. To make way for the added production, Teradyne leased another floor at its 87 Summer Street headquarters.

There is scant evidence that the test and measurement establishment was paying much attention to Teradyne in 1963. Hewlett-Packard, the clear leader in electronic instrumentation, had chosen its next frontier, and it was not semiconductor testing. H-P had instead set its sights on medical instruments, buying Massachusetts-based Sanborn Instruments. Tektronix was content to bolster its market leadership in oscilloscopes, and General Radio, based only 25 miles west of Boston in Concord, was also oblivious to the threat that Teradyne represented. In the years to come, all three would make their runs at the semiconductor test business, but by then Teradyne would have built and armed its fortress.

DeWolf and d'Arbeloff were quite happy to grow their business out of the spotlight. Thus, Teradyne's progress was decidedly low-key, with very few conference papers or magazine articles of the type that would bring unwanted attention to the revolution in the making. The rest of the world would get wise soon enough; why invite competition now?

Teradyne did not, however, escape the attention of one established instrument maker. In 1963, the young Teradyne was visited by Ikuo Takeda, chief executive of Takeda-Riken, a Japanese company he had founded in 1954. Takeda had somehow heard about Teradyne's novel approach to testing and decided that it was worth a trip to Boston. He was graciously received and allowed to tour and photograph the Summer Street plant. Eventually, Takeda-Riken would change its name to Advantest and become a hugely successful company and Teradyne's most formidable competitor.

6

• • •

A Company Takes Shape

Each spring, ARD held a meeting for its investors at which it paraded the people and products behind its many ventures. Teradyne was on hand at the March 4, 1964 meeting to "show and tell," but the star of the meeting was Digital Equipment, which unveiled its PDP-8, the world's first commercial minicomputer. The Teradyne founders were intrigued. Their testers had been programmed, first by rotary switches and modules, then by "digiswitches," banks of thumb-wheel-operated switches that allowed DeWolf and company to keep pace with the growing number of tests demanded by transistor makers. An additional layer of programming complexity involved "binning," or the sorting of components on the basis of test results. A transistor that passed all its tests would be assigned to "Bin 1." A transistor that failed some tests but that might still be useful for certain applications would be placed in "Bin 2," and so on, down to the component that barely exhibited any gain but that might be suitable for use in toys. To program the sorting scheme, DeWolf devised a clever "bin computer," a front-panel matrix of holes into which one could plug magnetic pins to designate the tests necessary to qualify parts for the various bins.

The programming burden, in short, was growing, and Digital Equipment's new minicomputer might — just might — provide an answer. A trip to Maynard was in order. The visit was set up through a test engineer who, it happened, had been hired by Digital on d'Arbeloff's recommendation. At Digital, the engineer had assembled a computer-operated module tester, which he proudly demonstrated to DeWolf and d'Arbeloff during their visit.

"Hello, Nick and Alex," the tester's teletypewriter clattered, to prove its mettle. Then the operator selected one of four modules to be tested, and the PDP-8 swung into action, switching the appropriate instrumentation and sequencing the tests. It was a seminal moment in test history, but it did not impress DeWolf, who spent the return drive to Boston listing the shortcomings of Digital's kluge. His critique was no doubt right on the money, but d'Arbeloff, looking through a different lens, thought Digital's tester was one of the most exciting things he had ever seen.

DeWolf's criticisms notwithstanding, he was not about to dismiss the minicomputer out of hand. It offered an enormous amount of computing power for a modest price, and this power naturally resonated in the mind of the master technologist who, after all, had been among the first to sense the potential of semiconductors in computers. So DeWolf and d'Arbeloff plotted their next step: They would send two of their best engineers, Joe Hendriks and Harvey Goldenberg, to a programming course at Digital, with instructions to report back. Their verdict: The computer was interesting, but it wasn't suitable for their purposes. It didn't have enough memory.

The time had come to test the waters in Europe. Early in 1964, Abe Bluestone, formerly a Components Marketing Manager for Raytheon and now an independent European sales consultant, was engaged by d'Arbeloff to sell the Teradyne line from a base in Paris. Thus began a long relationship, which saw Bluestone become the Company's sales manager a few years later. Bluestone's tour with Raytheon had given him a long list of contacts in France, Germany, England, and Sweden, and he mined these contacts energetically, demonstrating various Teradyne testers at Thomson, Bull, IBM, COSEM, TI, SILEC, and virtually every one of the semiconductor manufacturers on the continent. He described these visits and demonstrations in excruciating detail in voluminous correspondence with d'Arbeloff, many of his letters consisting of seven or eight pages of single-spaced typing. These were no travelogues, but point-by-point technical analyses of each customer's situation and of the adaptability of the Teradyne equipment to its needs. Bluestone was a workaholic, and by his sheer brute force Teradyne began meaningfully penetrating the European market. Soon Teradyne, still a very small company, had its first overseas sales office, in the city of Alex d'Arbeloff's birth.

The sales force, now led by d'Arbeloff and Lucey in the United States and by Bluestone in Europe, continued to expand. Representatives were replaced by firms considered to have stronger personnel and more clout. One of these was Instrument Specialists, Inc. ("ISI") which was signed to cover the West Coast. Another was Ossman Instruments, another Hewlett-Packard castoff, in upstate New York.

Teradyne's first customers were manufacturers of semiconductor components, but by 1964 a new market was emerging: the users of these components. A few disastrous field failures was all that was needed to convince makers of electronic equipment to begin testing what they were buying, rather than relying on their suppliers' representations. An impressive array of inspection equipment could moreover be used as airtight evidence that a failed lot of components could be returned for credit.

Thus emerged the "incoming inspection" market, recognized by Teradyne in its 1964 Application Note entitled "The Economic Justification for 100% Inspection." (The justification extended only to transistors and diodes; not even the most zealous salesman could argue the case for resistors.) At first glance, the incoming inspection market appeared much more attractive than production-line testing. For every semiconductor maker, after all, there were thousands of semiconductor users. Teradyne would answer the siren's call — up to a point. DeWolf and d'Arbeloff would continue to design and produce their testers for production-line use; if incoming inspection departments wanted to buy the same products without modification, that was fine, but there would be no design concessions. The founders' instincts were, as usual, on target. Although the inspection market would briefly approach 20 percent of the total market for semiconductor testers, it never lived up to early expectations.

The business was now sound enough to allow the young Company the luxury of turning down business it thought unsuitable. Inquiries for high-power testers were routinely deflected, as were requests that ran afoul of DeWolf's "speed kills" canon. Some of the requests were relayed by Teradyne's man in Paris, who was trying to match Teradyne's "arrogant" standards with the idiosyncrasies of his European customers. Finally, d'Arbeloff fired off the following: "I don't like the idea of 'design your own tester.'... Every option is an unfortunate necessity rather than a happy addition." Then, in case Bluestone missed d'Arbeloff's drift, there was this: "Please forget about European specs. There is only one T207 and this is described in the latest brochure." Teradyne may have seemed to be churning out an endless variety of "specials," but in fact these were all circumscribed by a rigid set of design principles. Within the boundaries of these principles, Teradyne could be very flexible, but the principles themselves were as inviolable as the sealed modules inside the testers.

A project that consumed DeWolf for much of 1964 was the design of a family of reed-relay scanners, designed to be used as accessories for Teradyne testers. In a typical configuration, the scanner would be programmed (by a pin board) to scan a diode matrix for wanted or unwanted diodes. The heart of the scanner was its reed relays, and these were supplied by Gordos, a New Jersey manufacturer, to DeWolf's specifications, then burned in at Teradyne for several million operations, and then tested on relay testers designed and built by Teradyne. The result of all this was arguably the world's most reliable reed relay — and the world's most reliable scanner. But it turned out to be a bad fit in the Teradyne product line, and the scanner line would sell in only moderate quantities. On the other hand, the project gave Teradyne unparalleled experience with reed relays as test system components, and the Company was able to exploit this for many years, at one point even marketing a line of relay testers.

Teradyne's competition in the 1960s consisted mostly of the in-house test departments of the major semiconductor manufacturers, notably Fairchild and Texas Instruments. The offerings from both lacked the elegant touches of the Teradyne testers, such as an oscillation detector to keep transistors from acting up under measurement, but both were backed by their powerful parents and strong sales forces. Having a major semiconductor maker as parent was of course a two-edged sword, as most semiconductor makers were uncomfortable contributing to their competitors' profits and were even more uneasy over the creation of a channel through which vital technical information could flow to a competitor. In fact, the test arms of the semiconductor makers were scrupulously careful to avoid conflicts, but without doubt Teradyne's status as an independent company was a competitive plus.

The development of the semiconductor test industry was following a classic pattern: The component makers first made their own testers, out of necessity. Then, as the component business grew, these manufacturers sought to develop a new revenue stream around the testers they designed — testers of which they were understandably proud. Eventually, however, the best of the test equipment designers would realize, as had Nick DeWolf, that designing testers for a test equipment company had to be more rewarding than designing testers for a component company. But the test system industry (the term "ATE" was just beginning to gain currency) was still young in 1964, and the component makers offered enough excitement to keep their test engineers home on the range.

The test landscape at the time was discussed by d'Arbeloff in an October 30, 1964 letter to Bluestone:

"Motorola has gone into the test equipment business, but mainly in the IC field... There is no doubt that the IC business will be very large, and outfits like Motorola, Texas Instruments, and Fairchild will be in there pitching. However, Teradyne will still have the best product line, and I am confident that we will get a share of this business in time."

Sales in 1964 were $913,000, more than double those of 1963, and bookings exceeded $1.2 million. The year was ending, moreover, on a high note. Orders were pouring in for diode and resistor testers, the transistor testers were starting to sell, the sales network had been strengthened, and a beachhead was established in Europe. The Company was profitable, and potential investors were knocking at the door, among them one Captain Maurice DeWolf, now retired and living on the French Riviera. But, *trop mal*, there was no stock for sale.

The Board of Directors, in 1964, consisted of the two founders, Gene Lyne, and John Shane, who had replaced Bill Congleton as the ARD representative. DeWolf's title was President and CEO, and d'Arbeloff, though technically the Company's one and only Vice President, signed his letters as "Sales Manager," in the quaint hope that the recipient would judge Teradyne to be a larger enterprise. Gene Lyne, still with the family law firm, served as Secretary and his brother Kerry, also a lawyer, was elected Assistant Clerk. D'Arbeloff and Dick Lucey managed U.S. sales, Ed Reddy manufacturing, and Bill Attridge field support. Bluestone (still technically an independent representative) managed European sales. Hendriks was the Company' s Chief Engineer, and Joe Lewey supervised the final testing of modules. In fact, most of the new engineers hired by Teradyne were asked to begin their Teradyne careers in the Company's test department. There was no better way, in the eyes of the founders, to learn about the practical problems faced by Teradyne's customers.

A recurring problem for the Company in the early 60s was the lack of adequate documentation. A certain amount of nondisclosure was part of the Teradyne mystique, of course: Customers could not expect to receive detailed circuit descriptions and diagrams. But clearly written instruction manuals were a must, as were applications notes, catalogs, and other promotional material. The founders produced an astonishing amount of such material on their own, and the general literacy and even wit of some of the early pieces is noteworthy. And d'Arbeloff's mountainous correspondence of the period shows remarkable facility with the language he had come to late. But DeWolf and d'Arbeloff had much better things to do than write brochures. Into the breach stepped their old MIT classmate, Kemon Taschioglou.

Kemon, after graduating from MIT, served in the Air Force, graduated from the Harvard Business School, and spent seven years at Polaroid, becoming that Company's Industrial Sales Promotion Manager. Like DeWolf, he was boundlessly enthusiastic about whatever he was working on at the moment. Like d'Arbeloff, he had a volatile personality, subject to wild mood swings. Most relevantly, he had a sure sense of what words, what designs, what colors were exactly right for a given booklet or trade show booth. In the decades to come, Teradyne's image as a "class act" would be attributed in part to the consistently high quality of its promotional materials. This standard was set by Kemon Taschioglou; all his successors had to do was follow his tracks.

Taschioglou's first assignment on arriving at Teradyne was the creation of a product catalog. The ingredients were choice: a number of DeWolf aphorisms that expressed the Company's philosophy of testing,[1] and an array of exquisite photographs,

[1]Some examples:

"Eternal vigilance is the price paid for a single adjustment point."

"An automatic component test instrument hardly makes sense unless the training course which teaches how to program it takes no more time than the average coffee break."

"The final cost of the one bad component that gets into an assembly often is greater than the purchase cost of the entire lot from which it came."

"If your harried inspector has confidence in the consistent accuracy of the test instrument, he'll be willing to put his foot next to the spike when he swings."

most of them taken by DeWolf, showing the legions of Teradyne testers which, like the army of Xian, were created just for show. (But if you ordered one of these soldiers, it would march to battle.) The catalog was a handsome piece and it was distinctively Teradyne. It had an attitude one would not find in the catalogs of other test equipment companies.

7

• • •

Harnessing the Computer

Ever since the 1964 ARD meeting, DeWolf and d'Arbeloff could not shake Digital Equipment's PDP-8 from their minds. Here was a full-bodied computer, adaptable to a virtually limitless number of applications, for $18,000, a small fraction of the cost of anything even approaching its power. There had to be an important place for such a computer in the test equipment universe. Though not sure exactly where that place was, Teradyne in late 1964 quoted a T231 Transistor Tester plus the electrical interconnection to a PDP-8 — but with no software — to a Transitron engineer named Milt Collins. Before landing at Transitron, Collins had been Manager of Advanced Circuit Development at Olivetti. Earlier, during a seven-year stint at IBM, he had been Manager of Basic Circuits for the Sage computer. Now, after four years at Transitron, his portfolio included expertise with both computers and semiconductors, and he could see the potential for synergy.

It was clear to DeWolf and d'Arbeloff that Collins had exactly what they needed, and he was sure he could program the PDP-8 to operate the Teradyne tester. In February 1965, he joined Teradyne to make good on his boast. He and Kemon Taschioglou began work at Summer Street on the same day.

Milt Collins deserves full credit for directing Teradyne's pioneering efforts in computer-controlled testing. He had a quick mind and an ability to focus on any problem until it was brought to heel. His somewhat flinty personality might have proved a career handicap elsewhere, but DeWolf and d'Arbeloff were in the market for ability, not charm. The eye for discerning the real talent of prospective employees, without regard to their résumés or their social skills, was one of the most important talents possessed by DeWolf, and especially by d'Arbeloff, who was doing most of the hiring.

1965 began auspiciously. The Company entered the year with a large backlog, and Motorola, RCA, and others were lined up to buy the computer-operated systems Teradyne was offering — somewhat optimistically, given the fact that Milt Collins was barely on board and that Digital was quoting six months for delivery of a PDP-8. D'Arbeloff ordered two of these in February, two more in March, but was forced to suggest that customers wanting early delivery of their new test systems substitute the PDP-5, at a price of $27,000 vs the PDP-8's $18,000. (Customers who were in a position to pressure Digital for faster delivery of their own PDP-8s were of course free to do so, and some did.)

When Collins arrived at Summer Street, he hit the ground running, and before long Teradyne's T231 transistor tester cum PDP-8 could be quoted complete with a master operating program ("MOP") developed by Collins.

What DeWolf and d'Arbeloff foresaw, albeit dimly at the time, was that a minicomputer offered much more than a programming medium. The interaction between computer and test

Captain Maurice DeWolf, USN.

Marian Watts DeWolf.

Nick DeWolf on the beach, Bay Head, NJ.

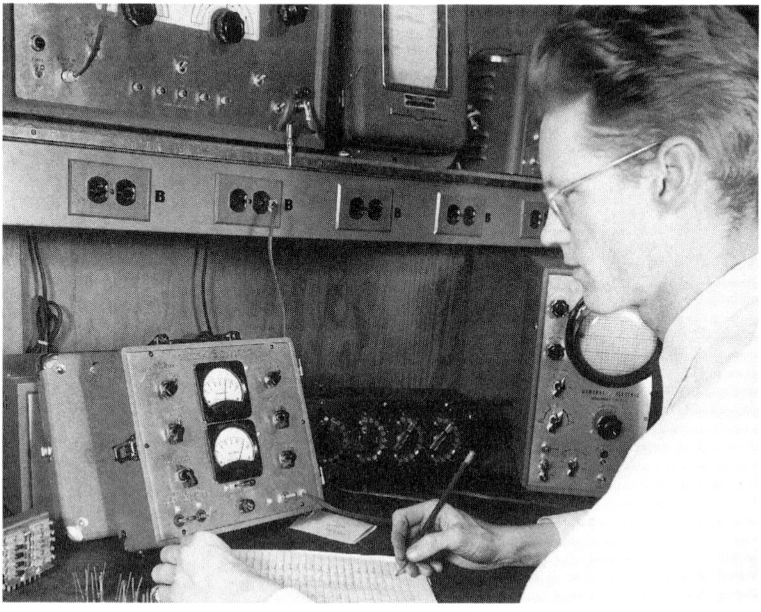

Nick DeWolf and his Germanium Diode Checker, at General Electric.

The Baron and Baroness Tiepolt, in Paris.

Alex (Sasha) and Vladimir d'Arbeloff as young men.

Vlad and Catherine d'Arbeloff and their sons Dimitri (Dee) and Alex, on the deck of the Europa, en route to New York, December, 1936.

The three d'Arbeloff brothers — Alex, George, and Dee — at home in Connecticut, 1952.

Teradyne's first home, on the third floor of 87 Summer Street, Boston.

The plant at 183 Essex Street, Teradyne's headquarters from 1965 through 1984.

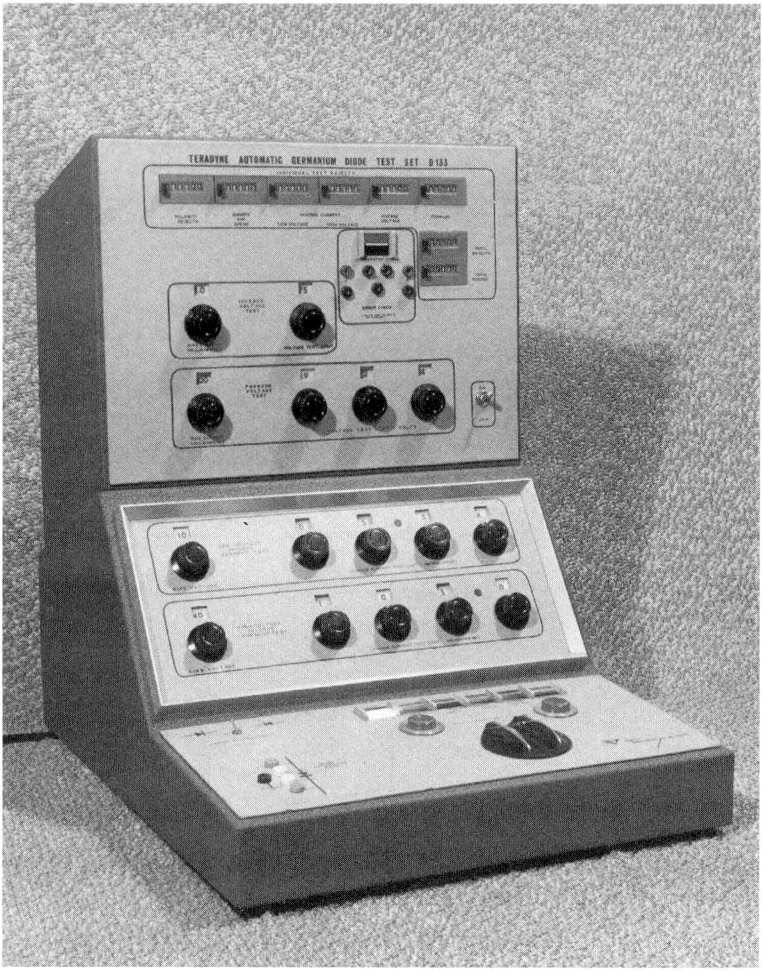

Teradyne's first product, the D133 Diode Tester. Test limits were programmed by front-panel rotary switches, diodes were placed on a magnetic test clip, the operator pushed a right- or left-hand "start" button, and test results were indicated by green (pass) or red (fail) lights.

The Company founders on the production floor at Summer Street.

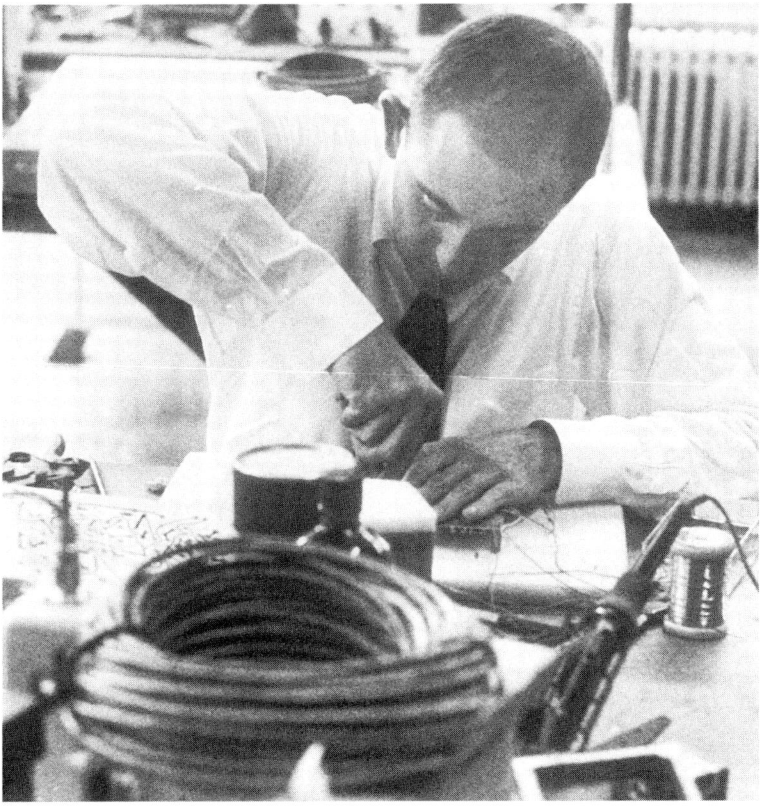

Alex d'Arbeloff, soldering at Summer Street.

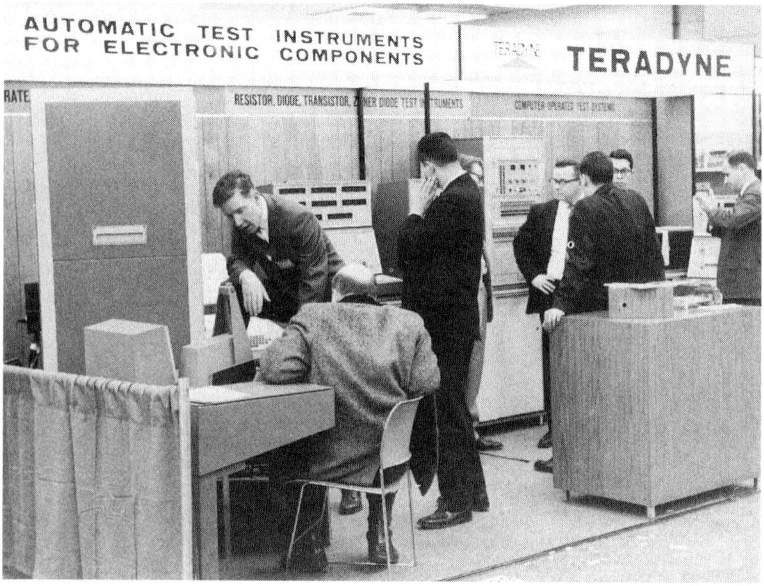

The booth at the 1967 IEEE Convention in New York. Stu Peters is at far left, and Dick Lucey is to the right, hand on hip.

Abe Bluestone and Milt Collins.

instrumentation opened up a sea of possibilities. The computer could initiate a measurement, for instance, and then, depending on the result of that measurement, decide what measurement to make next. With the computer, for instance, the process of successive approximation, in which the tester quickly homes in on a measured value by running a series of comparisons against converging values, was easy to program and execute.

Computer-operated transistor testers were bought by a number of manufacturers, including Sprague, General Electric, TRW, RCA, ITT, and Amperex. The experiences were not without trauma. Dick Lucey remembers selling one of the first units to ITT, in Florida. "It took a half hour to load the MOP," he recalls, "and the PDP-8 went down often. Even so, it was very productive."

At least ITT's Director of Quality Assurance thought it was productive. This was an ex-Bostonian and a former Transitron employee named Dick Bailey, and while Lucey was selling the T231 he was also selling the excitement of working for Teradyne. Bailey was soon recruited and brought to Boston, where he joined Lucey and another ex-Transitron colleague, Stu Peters, in sales. Bailey, with no prior experience in selling, would become one of Teradyne's all-time great salesmen and something of a legend in his own time.

Teradyne had established a reputation for building productive testers, but the PDP-8 brought new meaning to the word. In the view of DeWolf, d'Arbeloff and company, the case for computerized testing was now airtight. The competition, meanwhile, kept telling themselves that the computer was a gimmick, ultimately to be undone by its lack of memory. While they scoffed, Collins was grinding out thousands of lines of code.

Frank Burge, then a sales manager for Fairchild, later a trade magazine editor and columnist, recalls the competition's reaction to Teradyne's computer-operated testers:

"The DeWolf gang developed a tester with a PDP-8, and they started kicking our butts. Every time we had a lost-business meeting the sales guys would whine that we lost the order because we didn't have a computer. The engineering guys would wave their arms and claim the customer didn't really need a computer and if we had better salesmen we wouldn't lose the business."

Early in 1965, Bluestone's status changed from representative to full-time Teradyne employee. The d'Arbeloff-Bluestone traffic, in volume rivaling the Roosevelt-Churchill correspondence of World War 2, continued unabated, with Bluestone (Churchill) usually pushing for some advantage and d'Arbeloff (FDR) balking but eventually giving in, except where principle interfered. As soon as Bluestone became Teradyne's official European Sales Manager, he began scouting the region for representatives. Meanwhile, DeWolf and d'Arbeloff represented the Company at a Paris trade show in April, and Joe Hendriks visited customers in Europe during a summer trip to his native Holland.

The level of Teradyne's activity in Europe was totally out of scale, considering the fact that the Company's total worldwide sales were less than $1 million in 1964. At U.S. companies 10 or 20 times Teradyne's size, Europe was out of bounds except for an occasional junket to Paris by the President. Why, then, was Teradyne expending so much effort in Europe?

First, because in Abe Bluestone Teradyne had found a human whirlwind who "knew the territory." Second, because although the prospects were fewer (Philips, Siemens, Thomson, Plessey, etc.), these were racing to catch the semiconductor wave and eager to buy capital equipment. Third, because U.S. semiconductor manufacturers were establishing plants on the continent. Fourth, and most important, because d'Arbeloff was a true internationalist, not a Bostonian who thought west was Worcester and east was Maine. The same attitude would see Teradyne establish itself in Japan long before that country became a semiconductor titan. To d'Arbeloff, it was all one world, which happened to be broken up into sales regions and time zones. Occasionally, there were also family connections that could be tapped, as when Uncle Sasha, living in London, found suitable office space for the Company in nearby Chertsey. DeWolf, too, was comfortable in expanding Teradyne's horizons. His parents, after all, had spent more time abroad than they had in the United States.

Teradyne's leap into Europe early in its history would prove profitable, not just on the income statement. It also led to the hiring of a number of Europeans who would join Teradyne's U.S. operations, in some cases becoming key officers of the Company.

In the fall of 1965, another experienced semiconductor salesman was hired. Byron Brooks had been selling transistors for RCA for several years, and one of his favorite customers was Nick DeWolf. "One night in 1961 I was sitting in my office when Nick called. He had just heard about a new transistor RCA had introduced, and he was hoping to get a few samples. I drove to downtown Boston that very night, and Nick and I had dinner — hot dogs at Joe & Nemo's."

Brooks's extra effort made an impression on DeWolf, and in 1965, when Brooks, then selling for RCA in California, showed up at Teradyne's Wescon booth in Los Angeles, DeWolf suggested that he join Teradyne. An interview with d'Arbeloff followed, and soon Brooks was managing west coast sales for Teradyne. "West coast" in this context meant anything west of Pittsburgh. The area east of that city was covered by Dick Lucey.

8

• • •

Making It Downtown

As the Company's production expanded to keep pace with booming orders, it was clear that the Summer Street plant could no longer contain the growing Teradyne. But the founders were as determined as ever to remain in downtown Boston. DeWolf still lived on Beacon Hill (and walked to work), while d'Arbeloff lived in Brookline, a 15-minute drive from the plant. So the search began for new space in the downtown area.

Just a few hundred yards from the Summer Street plant lay Boston's former leather district. Here, in an area of six square blocks, were a dozen or more underutilized buildings, mostly six- to eight-story, constructed of stone or brick. At one time, the area had been filled with thriving leather and woolen merchants, but most of these had left years ago, and the buildings were now empty or occupied by the studios of struggling artists and designers. One of the more imposing structures in the area was an eight-story brick building at 183 Essex Street, only a block from Boston's major rail terminal, South Station. Learning that it might be available, d'Arbeloff began negotiating in June, 1965, and was soon able to lease the top two floors of the 165,000-square-foot building for a dollar a square foot, with options to lease additional floors as necessary. Soon, work began on the necessary improvements to the top two floors of the building, which had slipped into disrepair.

The move to Essex Street in 1965 was a major event in the Company's history. Although the Company initially occupied only the top two floors of the eight-story building, the building was now effectively theirs, and potential recruits could be taken to lunch in downtown Boston and then walked down Summer Street to be shown the rather impressive structure.

The building offered startling contrasts to the visitor. On several floors, leather tradesmen toiled in dingy surroundings illuminated by bare light bulbs in sockets dangling from the ceiling. They were typically men in their 60s and 70s, wearing yesterday's shirts, suspenders, and sometimes even eye shades. Sharing the two creaking elevators with them as they went to and from their work were Teradyne's engineers, young and bright-eyed and well dressed. The old order was changing in Boston, and one could see it happening at 183 Essex Street.

At street level, large showroom windows revealed stacks of hides, under the sign "Shrut and Ashe." For several years, until Teradyne took over the entire building, this last vestige of the leather district's glory days greeted all visitors to Teradyne, which was now, according to Nick DeWolf, "the largest electronics company in the leather district of downtown Boston."

The pattern — new technology saving an architectural relic of a dying industry — had been set 25 miles west of Essex Street, where a thriving Digital Equipment had taken over the old woolen mills that lined the banks of the Assabet River. There, the enormous brick piles that had long been considered an eyesore were suddenly in fashion, and old mill cities like Lowell and Lawrence began dreaming of a similar renaissance.

Although the similarities between Digital's old woolen mill and Teradyne's old leather building were clear, Teradyne could not be satisfied with a Boston version of the Digital headquarters, which, though attractive, were decorated with more practicality than imagination. Looking for something different, Teradyne's promotional chief, Kemon Taschioglou, hired a designer named Bill Bagnell, noted for his flamboyant use of colors. Bagnell sized up the maze of pipes and ducts and conduits hanging from the Essex Street ceilings and decided against trying to hide them. Instead, they would be color-coded, in carefully chosen shades of purple, mustard, and orange. The garish color scheme was submitted to DeWolf, who of course thought it was sensational.

A room near the northeast corner of the eighth floor was chosen as the primary conference room. A huge oval oak table was installed, and attention turned to the room's chief architectural feature, a long, high wall, surfaced in drab plaster. Taschioglou poked through the plaster to find a brick wall waiting to be exposed. But his proposal to strip the wall was dismissed as too costly (a veneer of fake brick was suggested), and so, acting completely in character, Kemon, accompanied by his wife Rhoda, took hammer and chisel to the offending plaster on a Saturday morning to bring forth a handsome brick wall, which elicited compliments for years to come.

Floor by floor, Teradyne set about remaking the old building in its own image, while the soon-to-be-dispossessed leather merchants looked on in wonder. One morning, the Shrut half of Shrut and Ashe accosted d'Arbeloff as the latter entered the lobby.

"D'Arbeloff," the older man began, "I see you spending a lot of money on this building."

"Yes, we are," agreed the young entrepreneur.

"You know, I've been a businessman for many years, and let me tell you, it's not good for the money to be going out all the time. The important thing is for money to be coming in."

"Mr. Shrut," replied an indulgent d'Arbeloff, "I can assure you that there's lots of money coming in."

Teradyne was born in Summer Street, but it was at Essex Street that the venture became a company of substance and character. And it was at Essex Street that the philosophy of management conceived by its two founders gradually grew into a personality that was readily apparent to employees, customers, and suppliers. History shows that most great companies are distinguished by unique cultures imprinted by their founders. If the imprint is deep enough, it will survive their eventual departure.

Culture and substance: Both are necessary. It is presumptuous for the newborn company to talk about its culture. First it must prove itself in the marketplace; only then can it look inward and decide what kind of animal it wants to be. By 1965, Teradyne had proved itself. Its sales were doubling every year, it was highly profitable, and the market had recognized its important technical contributions. Our history has reached a point at which an examination of the Company's culture is in order. In no particular order, then, these were the rules Teradyne lived by:

Technical excellence was a given, as it would have been in any electronics company, but with a difference: Technical achievement was not an end in itself, but a means to an end, which was the economic enrichment of the customer. Thus, while other companies may have pursued the extra digit of resolution or the extra hundredth of a percent of accuracy,

Teradyne pursued reliability and throughput. The pursuit in fact led to many important technical contributions, but they were incidental to the main objective.

The focus was always on the real business at hand — designing, making, and selling products. All else was subordinated. Believing that titles and organization charts might distract employees from their main tasks, the founders outlawed them. Business cards bore the employee's name and the Company's address and telephone number, period. As for organization charts, the story made the rounds of the engineer who, a few months after being hired, complained to d'Arbeloff that he didn't know whom he reported to.

"Do you know what it is you're supposed to be doing?" d'Arbeloff asked him.

"Of course I know that," answered the engineer.

"Well then," d'Arbeloff replied, "What difference does it make whom you report to?"[1]

Egalitarianism was very much in style in the California electronics companies that were springing up, but the perkless style at Teradyne was fairly unusual among Boston-area companies at the time. Private offices with doors were out, doorless cubicles were in. There was no separate dining area for executives. All air travel was coach-class.

Business integrity permitted no ethical breaches whatsoever, and any transgression was cause for immediate termination. Screenwriters who insist on stereotyping business people as venal should have studied Teradyne, where straight dealing with customers, suppliers, and, above all, with other employees was a sacred principle.

[1]Most people probably take this story with a grain of salt, but Dick Bailey, a Teradyne salesman for 30 years, insists "most of the time I was at Teradyne I never knew who I worked for."

A byproduct of the above was openness and easy communication throughout the Company. Anyone could sail into any office, including the founders', at any time. Complaints were often aired with brutal candor, and yelling was not uncommon. But nothing was taken personally, and once the issue was settled, life went on as before.

Another byproduct of openness was open recruiting. Thus, if one wanted to hire a technician or secretary away from another Teradyne employee, it was up to the latter to defend his turf. And in fact, there has long been a steady flow of people from one part of Teradyne to another, and from one function to another.

The founders also took great pains to ensure that what they referred to as "individual contributors" were rewarded no less than were those whose skills lay in managing others. A topnotch engineer, they felt, should not feel compelled to become an engineering manager in order to do well financially. In fact, this principle served the company well, as many who lacked management skills were able to enjoy successful careers at Teradyne while making important contributions to the Company's growth.

None of these principles was written down at the time, and in fact the founders resisted the very notion of issuing anything that smacked of an edict. Common sense was to be the rule, and common sense would lead inevitably to openness, candor, integrity, and all the rest. There was also a pragmatic streak, which dictated that the Company would not try to change the world, but would deal with society as it was. This attitude was tested during the 1960s, when the use of illegal drugs penetrated the inner city, Teradyne included. After considering the matter, the founders decided against publishing a memo expressly forbidding drugs on the premises. Yet no one doubted that anyone found to be trafficking in drugs would be gone in an instant.

Some years later, in the 1970s, the principles that guided Teradyne's behavior were codified in a pamphlet that came to be known as "the green book" because it was printed on green paper. It drew a mixed response — validating the founders' earlier belief that such things lose effect when they are stated explicitly — but it serves as a valuable window through which one can view the essential culture of the Company and is thus included in an Appendix.

With the added motivation to expand at Essex Street and create "the Teradyne Building," DeWolf and d'Arbeloff poured it on in the mid-sixties. The framework of a management staff took shape. Dick Lucey and Byron Brooks managed domestic sales, and in Paris Abe Bluestone, preparing to move to Boston, recruited his successor, an Englishman named Tony Wakeling. Milt Collins was in charge of software development, assisted by Norm Turner and, in the summer, by a young Harvard undergraduate named George d'Arbeloff. Gene Lyne, one of the Company's four directors and its Counsel and Clerk from the start, left the family law firm to become full-time Treasurer. Kemon Taschioglou ran sales promotion, and Field Service was headed by yet another ex-Transitroner, Stu Peters, assisted by Joe Lewey. Frank Hazelwood was hired as Controller. Joe Hendriks and Bill Attridge were senior engineers.

9

• • •

The Blockbuster J259

Sales doubled in 1965, and they doubled again in 1966, from $1.9 to $3.8 million. Articles about Teradyne and its founders began appearing in the Boston newspapers, job applications poured in, and investors asked in vain for copies of the Annual Report. The "go-go years" of the stock market were in full flower, and the investor banking community was looking for promising companies like Teradyne.

Having converted all its preferred stock into common in 1965, the Company had a total outstanding capital stock consisting of 40,110 shares of common, including 1000 shares in the treasury. Since it was highly profitable and had ready access to debt capital from ARD and others, Teradyne had no need to float an equity offering, and both DeWolf and d'Arbeloff preferred to wait until the Company was much larger before taking the Company public.

Teradyne's rapid sales growth was due in large part to the ability of its founders to sense what customers needed (which, held DeWolf, was rarely the same as what they wanted), but it also reflected the coming of age of the semiconductor industry. The contest between Route 128 and San Francisco's Bay Area was essentially over, and talent began flowing from the Sylvanias and Spragues and Transitrons of the Boston area to California's Fairchild Semiconductor and thence to its many spin-offs. Not that Massachusetts was suffering; Digital Equipment was becoming a computer giant, Raytheon a major military contractor. Among the smaller companies, many, including Teradyne and Analog Devices, showed great promise.

In the semiconductor industry, the spotlight shone on the integrated circuit, which promised to remake the landscape of all electronics. Just as the transistor had raised the testing bar by requiring more tests than did diodes, the IC would require so many tests that programming was sure to be an issue. To DeWolf, it seemed a heaven-sent opportunity to prove the minicomputer's value as a test controller, and he knuckled down to the task of designing an IC tester.

Teradyne's first integrated circuit tester began life as the J229. (The letter prefixes used to date would have dictated an "I" for integrated circuit, but that was considered confusing, so the next letter in the alphabet was chosen.) Its price, complete with PDP-8, was $58,500. Initially, the system was designed with discrete components, since DeWolf was not ready to trust the new integrated circuits to deliver Teradyne-level reliability. As it became clear that the IC in fact made systems more, not less, reliable (there were fewer of them to fail), DeWolf, aided by a young engineer named Peter Konde, redesigned the entire system using ICs. The redesigned system was christened the J259.[1]

The J259, which combined the hardware reliability for which Teradyne had become justly famous with the software

[1]An inadvertent breach of DeWolf's prime-number rule. DeWolf had chosen 257, which is prime, but brochure-writer Kemon Taschioglou changed it to 259 to reflect a design change. DeWolf was not amused, but the brochures were in print and the name stuck.

wizardry of Milt Collins, was a sensation almost immediately. The first systems were sold to Boston-area IC producers Sylvania and Sprague,[1] and the key west-coast breakthrough occurred when National Semiconductor purchased the system. National, whose Production Manager was another of the ubiquitous Transitron alumni, became an important California reference for other companies interested in the revolutionary new system — much to the discomfort of Teradyne's chief competitor, Fairchild Semiconductor, which continued to dismiss computer control as a gimmick. In time, even Fairchild's own semiconductor plant in South Portland, Maine would buy the J259 — on condition that the sale not be publicized.

The J259 consisted of a relay rack full of power supplies, an ingenious measurement system, a crosspoint matrix with banks of panel lights useful in troubleshooting, and, sitting adjacent to it, the PDP-8 and its paper-tape reader and teletypewriter. Its $58,000 price tag was more than reasonable in terms of its performance, yet DeWolf feared that anyone looking inside the cabinets would conclude that there wasn't $58,000 worth of "stuff" there. Of course, the key was the software, but the world was not yet accustomed to paying large sums for something that couldn't be seen or touched. A new Volkswagen, DeWolf noted, could be bought for $2000. Would people be willing to pay the equivalent of 29 Volkswagens for a J259? The answer was a resounding "yes."

Eventually, more than 150 J259s would be shipped, driving Teradyne's sales from $3.8 million in 1966 to $5.6 million in 1967 and $9.1 million in 1968. Teradyne with a single bold stroke had become the world's leading maker of IC test systems.

[1]Bill Russell, then a J259 user at Sprague and later a Teradyne software engineer for over three decades, remembers evaluating the system versus the competitors: "Compared with the 259, the others were just junk; there was no comparison."

The catalog still listed diode testers, transistor testers, resistor testers, and scanners, but the J259 was reshaping the Company in ways that could be only dimly perceived at the time. For instance, customers buying a computer-operated test system were required to send personnel to Teradyne in order to learn how to program the system, and soon a steady stream of test engineers came to Boston for the one-week courses, graduating with Teradyne diplomas to be framed and proudly mounted on their office walls. Experience with Teradyne's J259 was soon written into recruiting ads, and as the graduates migrated from one semiconductor company to another, they carried with them their Teradyne loyalty. Other courses in system maintenance were offered, and customer training became a major activity at Essex Street. Visiting students were berthed at downtown hotels, wined and dined at good Boston restaurants, and warned about the hazards of Boston's red light district at the other end of Essex Street.

The J259's measurement system, the M359, was a sterling example of DeWolf's art. Its function was to measure a current while forcing a given voltage, or to measure a voltage while forcing a given current, in either case guarding against extremes that could damage the IC under test. The elegance of its design prompted some companies, AT&T's Bell Labs among them, to attempt to buy the M359 as a separate item. No deal, said Teradyne; its business was systems, not components.

The impact of the J259 on Teradyne's history cannot be overstated. It was not simply the sales volume, which was considerable, or the number of new customers brought into the fold. The J259, which brought computer-operated testing to a level only hinted at in the transistor tester, opened up a whole new continent for Teradyne to explore. Clever engineers, both inside and outside the Company, began wondering what other testing feats might be accomplished by variations on the theme. In Chicago, engineers at Teradyne's midwest representative, Cozzens & Cudahy, began exploring the possibilities of using a modified J259 to trim film resistors to value. At Teradyne, engineers wondered whether the J259 platform might be useful in the testing of circuit boards and linear circuits. Others suggested designing a system that would extend the J259's capabilities to include dynamic testing of ICs (i.e., testing of a device's speed characteristics). These speculations unnerved DeWolf, who questioned whether the systems' capabilities could be extended without compromising reliability.

Milt Collins, meanwhile, was already thinking in terms of an encore and chafing under the programming constraints imposed by the PDP-8's 12-bit word length. Commercial alternatives to the PDP-8 were few at the time, and the idea of designing a Teradyne computer, featuring the "Mack Truck" ruggedness DeWolf demanded and the word length Collins wanted suddenly seemed attractive. DeWolf and Peter Konde thus set to work developing the M365, an innovative, intentionally limited computer that would control a number of highly successful Teradyne test systems in the following years. Eventually, the computer industry would offer Teradyne the features it needed for its test systems, but the M365 played a valuable role in bridging the gap at a critical time in the Company's history.

DeWolf thinks of the M365 as one of the first RISC (reduced instruction set) computers, and in a sense it was. More important, it was a monument to DeWolf's sense of engineering economy, being fabricated mostly out of NAND gates, flip-flops, and diodes. Core memory also permitted another DeWolf touch: plug-in circuit boards that could be swapped without loss of data. There were, needless to say, no fans or filters.

The Teradyne testing universe, in 1967, spanned integrated circuits, diodes, transistors, resistors, and Zener diodes. Capacitors were added when Teradyne acquired a small UK-based instrument company, Mark Instruments. Teradyne purchased the design of Mark's new automatic capacitance bridges and rechristened them as the Teradyne C302 and C357. The bridges would never become commercial successes, and Mark Instruments' claim to a place in this history rests on the fact that it brought to Teradyne several "Brits" who would rise to positions of prominence in the Company. Among these were Graham Miller, who would soon lead Teradyne's efforts in the linear IC test business, John Turner, an engineering manager who would eventually take over the linear IC test division and direct its resurgence in the 1980s, and Russell Ashdown, a gifted engineer responsible for Teradyne's "4TEL" telephone line test system. It was a rich harvest of talent, from an otherwise inconsequential acquisition.

In 1967, Teradyne purchased the Essex Street building for $900,000, financing the purchase mostly through a bank mortgage, partly with a $250,000 loan from ARD. The purchase agreement stipulated, in deference to the few remaining tenants (mostly leather merchants), that the building should not wear Teradyne signage until the Company occupied five of the eight floors. But that was beside the point; it was clear to all that Teradyne would soon take over the entire building, leaving the leather merchants to look elsewhere.

The building at 183 Essex Street was to serve as Teradyne's headquarters for more than 15 years, until the Company found another in-town derelict worth saving, and the $900,000 investment must rank as one of the better real estate deals of the time. Still, the building was not without its problems, some of them self-inflicted. There was, for instance, the matter of the flocked lobby. A Teradyne design consultant who was a passionate advocate of flocking (his car was flocked) sold management on the idea of electrostatically spraying orange flocking material on the front lobby walls and elevator doors.

Given the generally positive reactions to the interior color scheme that DeWolf had previously okayed, flocking probably seemed a reasonable next step. But the dirt that hung in the air in downtown Boston was attracted to the flocked surfaces like filings to a magnet, and the lobby soon assumed the look of a rundown brothel or of one of the seedy clubs at the other end of Essex Street. Worse, the flocking stubbornly resisted attempts to undo the damage, and eventually the elevator doors were replaced, the walls covered, and the flocking left for some future archaeologist to find and puzzle over.

The basement was the site of Teradyne's printed-circuit shop. Naturally, there were no windows, and the fumes defeated the best efforts of the exhaust system. Even more challenging were the effects of storm tides, since the Essex Street building, like everything else in the neighborhood, was built on a filled-in bay. Extremely high tides forced the storm sewers to back up, and at such times it was necessary to close the valves to keep the forces of nature at bay.

Naturally, none of this seemed particularly onerous at the time. In fact, it added spice to the adventure and may have helped hone Teradyne's talents for dealing with the industrial world. Teradyne systems were designed for use in the rough and tumble of the factory floor, with all the noise and dirt and abuse that goes with that, and it was only fitting that the systems came out of just such a gritty environment.

As Teradyne grew, it was obvious that, notwithstanding its other advantages, Essex Street was too small to accommodate the Company's ambitions, and in a few years at most another plant would be needed. The consensus in favor of remaining downtown was unshakable. DeWolf was still walking to work from his Beacon Hill home,[1] and d'Arbeloff was urban to the core. It was left only to find available space near Essex Street. A topographic map of the neighborhood was spread across the eighth-floor conference table by Gene Lyne, and he, d'Arbeloff, and DeWolf sized up the possibilities.

Fortunately for Teradyne, the success it had enjoyed as a downtown electronics manufacturer did not spawn a horde of imitators. Most electronics companies preferred the suburbs, where parking was not a problem and where most engineers and managers wanted to live. So the old leather district remained fallow, waiting for a Teradyne to bring it to life. Looking over the map, the trio noticed that one building dwarfed all others. It was located at 179 Lincoln Street, where it occupied an entire city block. Built around the turn of the century, it had originally been the headquarters of American Woolen Company. It had long been known as the Albany Building, because it had once been served by a siding of the Boston and Albany Railroad, before the area was remade to accommodate a highway. The building was only five stories high, but each of the stories offered a full acre of space. It was ideal, in all but one respect (an old bugaboo, parking), and negotiations for its purchase were set in motion.

The Board of Directors had remained unchanged for several years, consisting of the two founders plus Gene Lyne

[1]DeWolf's long love affair with Beacon Hill was starting to wane, however, as he became embroiled in controversy triggered by the arrival of drug dealers in the area.

and John Shane (ARD). In 1967 a fifth director was elected: Henry M. Watts, Jr., Nick DeWolf's uncle, who was an early investor in Teradyne and a recently retired member of the New York Stock Exchange's Board of Governors. Watts, who lived in Philadelphia, was Teradyne's first out of town director, and in recognition of this fact the Board authorized a Director's fee of $50 per meeting — for out-of-town directors only! Watts was to prove immensely helpful over a long tenure on Teradyne's Board.

The two founders took great pains to avoid the mistakes they had seen at their previous employers, especially with regard to ensuring that employees were able to share in the Company's success. Annual bonuses and yearly salary reviews were in place. Key employees were given stock options. Now, in 1967, DeWolf and d'Arbeloff allowed certain employees to buy 5000 shares of their personal holdings at a price of $25 a share. These were eagerly snapped up.

1968 was a banner year for Teradyne. The J259 was the talk of the industry, and semiconductor production managers begged Teradyne to accelerate production to satisfy the growing demand. One semiconductor producer made so bold as to write d'Arbeloff a scolding letter, warning that if Teradyne wanted to play in the big leagues it would have to learn how to ramp production faster. (In the sharp downturn that would hit in 1970, this company would be one of the first to cancel orders.) Personal relationships were invoked in the hope of moving up in the delivery queue. Best of all, the J259 had the market essentially all to itself, the competition still refusing to believe that computer-operated testers were more than a fad.

The Company's roaring success — and that of the semiconductor industry — stood in stark contrast to the mood of the world in 1968, one of the most turbulent years of the century. The Vietnam War, the assassinations of Martin Luther King and Robert Kennedy, student riots, the stormy Democratic convention in Chicago — all were shaking the country to its foundation. The semiconductor industry had its own convulsions, as Fairchild Semiconductor, easily the most exciting chip company of its era, was hit by a wave of defections, setting in motion a chain reaction that would produce a new generation of semiconductor manufacturers, the most noteworthy being Intel.

The year also marked Teradyne's first venture outside electronic testing, when it decided to back a trio of entrepreneurs who were forming a new company to produce backplanes. All three were out of Transitron: Windsor ("Windy") Hunter, who would manage the new company, had run Transitron's highly profitable diode operation, and Tom Neavitt had been Transitron's sales manager. The third member of the team, Len Johnson, was a skilled connector design engineer. (Transitron had entered the connector business via an acquisition.) The would-be entrepreneurs, seeking venture-capital funding, asked their old Transitron colleague DeWolf for help in securing introductions, but DeWolf, after reviewing the business plan, decided that Teradyne itself would finance the venture, then named Genex.

D'Arbeloff was cool to the idea, believing that Teradyne's first priority should be to build its core business without distraction. But once DeWolf's passions were in gear, argument was useless. Genex was acquired in late 1968 for 12,500 shares of Teradyne, plus an additional 12,500 shares contingent on earnings over the next two years. Teradyne's new division began as Teradyne Components, operating out of an old cookie factory in Lowell, Massachusetts. The initial product, designed by Johnson, was a clever implementation of the two-piece, fork and blade interconnection system specified by the Navy (NAFI). The new Company also decided at the outset to design and build most of its own fabrication and assembly equipment, and it thus was quickly able to add capabilities not available to competitors (e.g., the high-volume barrel-plating of Johnson's new fork and blade contacts) as well as a level of ruggedness and reliability that was totally in tune with the Teradyne theology.

A backplane is a connector panel that integrates the circuit boards in an electronic system, supplying both the mechanical support and the means of interconnecting the boards' electronics. Teradyne Components' attention was centered on the means of connecting the circuit boards to the backplane. The conventional method was to laminate gold fingers along the edge of each circuit board and to mount an array of mating, female connectors on the backplane. A more expensive, but far more reliable, two-piece system was favored by most military contractors. In this system, a separate connector was mounted on the circuit board, and this connector would plug into a mating connector on the backplane.

Teradyne Components was putting its eggs in the two-piece basket, featuring its new tuning fork connector. The first customer, Kearfott, awarded Teradyne an order for backplanes for a Navy program, and Teradyne Components was off and running, never to look back. In the years to come, this Teradyne division would become a major supplier of backplane systems used in military programs. Later, the expertise gained in the military market would turn into commercial gold when the era of very large-scale integration (VLSI) demanded the high-speed, high-density interconnection that only a two-piece system could provide.

D'Arbeloff's initial coolness notwithstanding, Teradyne's gamble on the backplane business proved to be by far the Company's most successful venture outside of semiconductor testing.

The launching of Teradyne Components, the acquisition of Mark (completed in 1968), the modernization of Essex Street, and the need for working capital to fund the Company's rapid growth strained finances somewhat, and early in 1969 a group of lenders, including ARD, Lehman Brothers, Greylock, and Old Colony, purchased $2.1 million worth of convertible debentures and warrants. Greylock's participation added a new member to Teradyne's Board: Bill Elfers, a Greylock founder.

With the J259 a runaway best seller, Teradyne was now attracting national attention as a new sensation in the testing business. Sales in 1968 shot up to $9.1 million, a 63-percent jump. After-tax profit was $650,000 (margins were squeezed by the cost of building a support infrastructure), and the balance sheet was solid. Yet the Company's success brought with it its own set of new challenges. The first, and most pressing, was the need to build a management team capable of transforming a successful small company into a successful mid-size company.

Related to this issue was the prospect of a public stock offering, necessary to attract talent and to help fund expansion. Employees, especially the handful who had been given stock options, began salivating over the prospect of an IPO. These were the "go-go" years of the stock market, when technology companies routinely were valued at 100 or more times their earnings. A reasonable projection of Teradyne's 1969 sales was $15 million. On that, the Company should earn well over $1 million. Times 100 equals $100 million. A few more calculations and the employee could dream of a new Mercedes.

There was a dark side. The new managers to be recruited would displace some of the major contributors to the Company's early growth. And no longer would Teradyne be the private domain of Nick and Alex. There would be outside investors — institutional investors, who might or might not agree with management's priorities. Nick, as President, would be expected to act presidential, a prospect that jangled his anti-establishment nerves. Nick was not worried about barbarians at the gate; he was worrying that "the suits" would take over Teradyne.

What he called "B-school types" were not DeWolf's favorite people. In fact, anyone who pursued advanced degrees was suspect, since that degree of self-indulgence revealed a lack of the impatience that, in DeWolf's view, was the mark of a true achiever. At Teradyne, DeWolf's battalion of achievers were all shirt-sleeve engineers. Some had engineering degrees, some did not, but they all knew what they were doing, they did it well, and that was all that counted. Teradyne engineers knew how to solder their own breadboards; they didn't expect technicians to do the grunt work for them. For that matter, they would sweep the floors and make the coffee if need be. This was the environment DeWolf thrived in, and the gentrification that lurked just over the horizon threatened all that.

DeWolf could rarely resist the chance to reveal his feelings on this subject. As the Company's public stock offering approached, an analyst from T. Rowe Price, a large financial institution, asked for a meeting with the Company's president. The meeting was scheduled, and the analyst began asking DeWolf a series of question about the technological aspects of automated testing.

"It's really not all that complicated," began DeWolf. "In fact, I could probably tell you everything there is to know about automated testing in a day, if you had the time."

The analyst, seeing an opening, bit.

"Actually," he said, "I have a doctorate in electrical engineering."

"Oh," answered DeWolf, "Well, in your case, maybe two days."

Another challenge facing the Company was the choice of the best marketing strategy to carry the success of the J259 into new testing territories. DeWolf, fascinated by operational amplifiers and other linear integrated circuits, saw opportunity knocking and spurred the development of a linear test system, the J263. At Cozzens and Cudahy, Warren Cozzens and Charles ("Chuck") Pearson began adapting the J259 to test motor controllers. Impressed with their ingenuity and with their obvious sales talents, Teradyne in late 1969 established a new organization, Teradyne Applied Systems ("TAS"), to market the J259 derivatives, which also included the W153 Circuit Card Test System (Teradyne's first board tester) and a system for trimming film resistors to value.

The J259 was, by design, a "slow" test system, which essentially measured the steady-state parameters — voltages and currents — of an integrated circuit. By avoiding high-frequency tests, DeWolf was able to design a system in the classic Teradyne mold: no fans, no filters, no adjustments, a 10-year warranty, service via plug-in replacement modules, and all the rest. Its slowness and its ruggedness were of a piece. Speed kills.

But now some key customers were demanding speed, or at least the ability to test, not only the IC's ability to move from a low-voltage state ("0") to a high-voltage state ("1"), but the amount of time it took to get there. Not everyone wanted dynamic testing, and it was arguable that one could verify a circuit's high-speed performance by means of dc tests, but the demand was there, and if Teradyne ignored it, its flank might be exposed to competitive attack. So the Company sought an entry that would not undermine the design principles so carefully nurtured at the Boston operation.

The vehicle was a joint venture with California-based E-H Research, a leading maker of pulse generators, run by an industry personality named Jack Hubbs. The new venture, called Triangle Systems, would give instrument maker E-H access to the semiconductor world and would give Teradyne entry into dynamic testing. To manage the venture, the partners chose Jack Salvador, a test veteran out of AAI Corporation, a supplier of testers to the military (the "AA" stood for "Aircraft Armament") now looking for a piece of the semiconductor test pie. Since E-H was located in Oakland, just across the Bay from what would soon become Silicon Valley, the parents of Triangle Systems favored that area. But Salvador held out for his beloved Southern California, and in 1969 Triangle Systems opened its doors at 17100 Ventura Boulevard, Encino, in the San Fernando Valley. Its President was Jack Salvador, its Treasurer Gene Lyne, and its Board included, in addition to these two, Nick DeWolf, Jack Hubbs, and E-H's Nathan Kessler.

Joint ventures are problematic under the best of circumstances, and this one soon came unglued. Within a year Teradyne bought out E-H's interest and renamed the operation Teradyne Dynamic Systems. Soon thereafter it was moved to larger quarters in nearby Chatsworth.

Back in Boston, meanwhile, DeWolf and his engineers were plotting a successor to the J259. This system would take full advantage of the integrated circuit's ability to compress hundreds or thousands of functions into a single package. It would also be the first to use the new Teradyne M365 computer. The old Digital Equipment paper tape reader was replaced by a magnetic tape unit (Cartri-File, from Tri Data), and the Teletype operator interface gave way to a new CRT terminal, the Teradyne M369. The system was named the J283, but in the Essex Street corridors it was known as the SLOT Machine, the acronym standing for Sequential LOgic Test.

The SLOT Machine would also be given a radically new packaging treatment, designed by Walter Kern, an inventive mechanical engineer whose passions were sports cars and outrageously bad jokes. The system elements were supported inside a skeleton made up of two vee-shaped aluminum extrusions, bolted together to form an "X" frame (looking down from above). This "kiosk" would be the standard Teradyne packaging system for many years, giving the Company's products a distinctive appearance while providing easy access to the system's innards from all directions.

The new linear IC tester received DeWolf's full attention, as he was firmly convinced that the world of operational amplifiers and converters would capture a major share of the semiconductor market. He and his team first created a J259 variation called the J261, then quickly followed up with the world's first stand-alone linear IC test system, the J263. Motorola and National Semiconductor were interested enough to order a few systems, but the market was slow to develop. Sharing DeWolf's enthusiasm for the linear IC tester,

Graham Miller began to spend less time on the capacitor testers he brought with him from Mark Instruments and more on learning how to market the J263.

There was more: Eyeing the incoming inspection market, DeWolf designed a compact, bench-top IC tester that could be manufactured and sold for one-tenth the price of a J259. The new tester, called the J133, was programmed by a pair of circuit boards. Having scored some success with the name "SLOT Machine," DeWolf was eager to find an equally catchy handle for the J133, and the new tester was christened "ACT I" for Analogic Circuit Test Instrument. Hopes were high for the new product, which was unveiled with much fanfare at an "ACT I" New York press conference, complete with press kit featuring a cover photo of a seductive "Eve" holding an apple (presumably moments before Act I).

Although a reasonable number of these testers were sold, it failed to live up to Teradyne's hopes. Companies that were serious about incoming inspection wanted system performance comparable to that of the testers used by their suppliers, so that test results could be correlated. Moreover, the J259 had proved that the computer was the ideal programming vehicle for an IC tester, and no matter how many J133 program cards Teradyne could supply, there would never be enough to keep pace with the number of new devices created by the semiconductor industry. But the basic miscalculation — which was to be made by other tester companies in the years to come — was the notion that the world would rush to buy a tester that offered 80 percent of the performance of a full-featured system at a fraction of the price. Usually, the market was not interested in 80-percent systems at any price.

Cracks were starting to appear in the DeWolf-d'Arbeloff relationship. The underlying cause was probably the fact that d'Arbeloff found it difficult to work in a subordinate position, and, although DeWolf was quite happy to leave most of the Company-building to his partner, the engineering department — the soul of any high-tech company — was unequivocally under DeWolf's control. D'Arbeloff by this time knew more than enough about the business and the technology to begin second-guessing product decisions, and he began to question these decisions with increasing frequency.

Whether it was angst-produced or not cannot be known, but d'Arbeloff sprouted a bleeding ulcer in 1968 and was forced to the sidelines for six weeks, three of them at the Massachusetts General Hospital. Upon his return, he forced the issues that had been troubling him. High on this list, as he saw it, was the fact that DeWolf, increasingly immersed in Beacon Hill politics, was no longer running the Company with the same intensity he had shown in Teradyne's early years. The choice, although it may not have been explicitly stated, was clear enough to d'Arbeloff: One of the founders would have to assume full command, with all that implied in terms of commitment. No substantive action resulted from d'Arbeloff's declaration, though he was shortly thereafter named Executive Vice President.

Largely on the strength of the J259, sales again soared in 1969, to more than $15 million. Profits grew even more, crossing the million-dollar threshold and virtually guaranteeing a successful public stock offering in 1970. With that bonanza in prospect, DeWolf and d'Arbeloff moved to make sure that key employees shared in the windfall, each offering about 10 percent of his holdings in a 1969 secondary.

The Company's five-year track record from 1965 through 1969 proved beyond doubt that Teradyne was on a roll. The numbers:

Year	Sales	Net Profit
1965	$1,904,000	$236,000
1966	3.810,000	446,000
1967	5,581,000	579,000
1968	9,112,000	650,000
1969	15,383,000	1,185,000

10

• • •

D'Arbeloff Recruits His Team

While DeWolf busied himself with the design of the SLOT Machine, the J133, and the new linear IC tester, d'Arbeloff embarked on a search for the managerial talent needed to carry the Company to its next stage of growth. Teradyne's products were excellent; of that there could be no question. But a company with staying power was yet to be built around those products, and d'Arbeloff saw that he would need experienced executives to help with the building. In his search, he depended chiefly on personal recommendations and contacts rather than professional recruiters. He did not share his partner's disdain for "business school types," nor did he insist on MBA degrees. In his view, only one qualification counted: The prospect had to have already proven his or her ability to do what Teradyne needed to be done. Of course, the talent had also to be convinced to cast his or her lot with Teradyne, but d'Arbeloff was a master salesman.

One can readily understand the sense of urgency that must have driven d'Arbeloff in his quest for experienced managers. Teradyne was at the time juggling too many balls for its small staff to handle. Teradyne Components, Teradyne Applied Systems, and Teradyne Dynamic Systems were all crying for attention. Since Byron Brooks and Dick Lucey were now product managers (for IC test and discrete semiconductor test, respectively), the U.S. sales network needed attention, and thus Abe Bluestone was summoned from Europe and Tony Wakeling installed in his place in Paris. Alex's younger brother George, having served his apprenticeship writing programs under Milt Collins and having received his degree from Harvard, was dispatched to Japan to help the Company's representative, Midoriya Associates, develop that market. The Essex Street building was filling up and the Lincoln Street plant would have to be brought on line soon. The coming public stock offering would consume an enormous amount of management resources. The breadth of issues facing the Company was staggering, and a few missteps could undo all that had been won since 1960.

Thus, in the two-year period from mid-1968 through mid-1970, d'Arbeloff worked feverishly to identify and recruit a new team of senior managers. His efforts paid off beyond his wildest dreams. The team he assembled served as the core of Teradyne's management for the next three decades, a period that saw the Company's sales grow a hundredfold. The group was noteworthy on several counts: As individuals, each possessed a superior intellect, impeccable integrity, and the ability and personality of a leader. As team players, they were able to work with each other and with d'Arbeloff without any discernible traces of ego or political infighting. Thirty years is a long time for a strong executive to work closely with other equally strong executives, and it is probably not surprising that most high-technology companies are plagued by frequent management turnover. Teradyne for the most part escaped this distraction, and credit for that fact properly belongs to the man who found these gifted people, attracted them to Teradyne, and led them for the next 27 years.

The bumper crops of 1968, 1969, and 1970 included a number of talented individuals who played major roles in the subsequent success of the Company. As mentioned earlier, the acquisition of Mark Instruments gave Teradyne three major-league talents: Graham Miller, John Turner, and Russell Ashdown. The fact that all three would eventually leave Teradyne in mid-career does not diminish the enormous contributions they made to the Company.

Others to join the Company in that era included Tage Peter ("Pete") Sylvan, Teradyne's best-credentialed engineer to that point. Sylvan, recruited from General Electric, was one of those electronic engineers who, like DeWolf, understood the way electrons behaved in a circuit and knew how to bend them to his will. A Fellow of the Institute of Electrical and Electronics Engineers, Sylvan was a quiet, unassuming worker whose legacy would include not only a long list of original contributions (among them an ingenious method of "daisy-chaining" a wired backplane so that it could be quickly tested) but a passion for organizing material in the Company's technical library, which he founded. The period also marked the hiring of Dennis O'Connell, an effective manager who had overseen Transitron's operations in Europe and would one day become Teradyne's Sales Vice President. And in early 1969 the Company hired a young production engineer away from the Foxboro Company to help upgrade its fast-growing manufacturing operations. His name was George Chamillard, and he was introduced to Teradyne by several classmates who were working for the Company and extolling its virtues. Chamillard turned down his first offer, but eventually d'Arbeloff convinced him that Teradyne was a good career bet. In the event, it was a better bet than either dreamed at the time, for Chamillard would one day succeed d'Arbeloff to become Teradyne's third president.

I should add here that in 1968 Alex's three-year-long campaign to persuade me to leave General Radio for Teradyne succeeded, and my own long association with Teradyne began.

The prospect of a public stock offering prompted the Company to shore up its banking and auditing resources. The Company's earliest banker had been Cambridge Savings, whose Gardner Bradlee had served DeWolf and d'Arbeloff well in the early years. Now, however, heavier artillery would be needed, and the First National Bank of Boston was brought on board. The auditing firm of J.K. Lasser was replaced by the larger Lybrand, Ross Brothers, and Montgomery, and Lybrand's Bill Siddall and Stu Osattin began the long process of restating results. (Both would later join Teradyne and become officers of the Company.) The first financial report bearing the Lybrand imprint would appear in 1969.

Many of those recruited in d'Arbeloff's 1968-71 talent hunt played major roles in building Teradyne, but two men in particular served as Teradyne's backbone and d'Arbeloff's right and left hand for the next quarter century: Jim Prestridge and Owen Robbins. Just as it is a truism that Teradyne could not have been born without DeWolf and d'Arbeloff, it is equally clear to most observers that Teradyne could not have achieved its present greatness without Prestridge and Robbins. The two merit closer attention.

James Arthur Prestridge was born in Jacksonville, Florida, on March 12, 1932. His father worked for a regional trade association representing the cypress industry and had come to Florida from Mississippi. His mother was a Georgian. He attended public schools in Jacksonville, and while in high school joined the Naval Reserve. This in turn led to a competitive appointment to the U.S. Naval Academy, from which he graduated in 1954 as a 2d lieutenant in the Marine Corps. There followed tours of duty at Quantico, Camp Pendleton, Newport (Naval Justice School), Guam, and Camp LeJeune. While on Guam he applied for admission to the Harvard Business School and entered the School immediately after separation in 1958.

Upon graduation in 1960, Prestridge entered Hewlett-Packard's training program at Palo Alto, later worked in a central sales support group, and, in 1963, when H-P was replacing its representative network with direct sales offices, moved to the new office in Rochester, NY. In 1965 he was transferred to the Rockaway Division in New Jersey (the old Boonton Radio, acquired by H-P) to become marketing manager for the Division's lines of instruments.

It was at about this time that Prestridge first met Alex d'Arbeloff. The occasion was the Wescon trade show in San Francisco, and the introduction was made at a bar one evening, by Austin Marx, an H-P marketer and a former classmate of d'Arbeloff's. Only pleasantries passed between the two, though d'Arbeloff, who had recently made a sales swing through upstate New York and who prided himself on "knowing the territory," was impressed to find that Prestridge knew more about that particular Teradyne territory than he did. This was exactly the sort of discovery that d'Arbeloff would file away for future reference.

In 1967 the two met again, at the NEREM trade show in Boston, where Teradyne was holding forth with its brand new J259. "I asked Alex if the thing worked," recalls Prestridge, "and, instead of giving me the usual song and dance, he smiled and said, 'Well, almost'. " D'Arbeloff's candor was exactly the kind of thing Prestridge would file away for future reference.

That same year H-P had moved Prestridge to the Boston area to head its new sales office — the successor to its long-time representative, Yewell Associates. In his new capacity as sales manager for H-P's New England region, Prestridge called on Teradyne in an effort to persuade the Company to buy one of H-P's new computers. The computer sale was not made, but in the process d'Arbeloff managed to convince Prestridge that what H-P really needed was a Teradyne transistor tester. (This sale was made.)

In 1969, H-P once again called on Prestridge to relocate, this time to Paramus, New Jersey to take over East Coast sales for its new line of computers. Although he dutifully bought a house in New Jersey and began readying his Concord, Massachusetts house for sale, Prestridge was hardly looking forward to another move. With perfect timing, d'Arbeloff offered Prestridge the job of building a sales organization for Teradyne, and Prestridge joined the Company on December 19, 1969.

Prestridge was unlike anything Teradyne had seen up to that point. Where d'Arbeloff was mercurial, Prestridge was even-tempered in the extreme. Where DeWolf was unabashed in expressing delight, Prestridge was always reserved. But what he shared with both DeWolf and d'Arbeloff was a first-rate mind. He was, moreover, far more organized than either, and a promise to "get back to you soon" was something you could take to the bank. Clearly, Teradyne's days of happy chaos were ending. The first of DeWolf's "B-School types" had arrived — in the opinion of many, just in time.

The sales landscape into which Prestridge sailed was in disarray. The representatives, whose sixth sense must have forewarned them of Teradyne's intentions to replace them, were restless. Abe Bluestone, the reigning sales manager, cooperated fully in ushering the new man around the Company, and d'Arbeloff (the de facto sales manager) and Prestridge huddled over countless dinners at Locke Ober's drafting the outlines of the sales organization they wanted to build. Finally, Prestridge officially took over as Teradyne's Sales Manager in the third quarter of 1970 — a quarter in which the Company's bookings actually went negative (i.e., cancellations exceeded new orders).

For the next three decades, Jim Prestridge's story is woven into the fabric of the Teradyne's history, and that is how we will tell it.

In March, 1970, only a few months after Prestridge's arrival on the Teradyne scene, another graduate of the Harvard Business School appeared: Owen Robbins, hired as Controller but obviously meant to become the Company's chief financial officer.

Robbins was born on November 18, 1929, in Pauls Valley, Oklahoma, the son of Welker and Jewel Patchell Robbins. His father was a one-time dentist, and both of Owen's brothers would become doctors. After schooling in Pauls Valley, Robbins entered DePauw University in Indiana, majoring in Economics and Political Science. He graduated in 1951 and promptly entered the Harvard Business School, receiving his MBA in 1953. He then served as an officer in the Army Finance Corps for two years, authoring the U.S. military's first study of business applications for computers. After discharge in 1955, he worked for Lybrand, Ross Brothers & Montgomery (later Coopers & Lybrand, now Price Waterhouse Coopers) for four years, then spent the next four years at General Electronics Laboratories, as Controller. In 1963 he moved to Providence, Rhode Island as Chief Financial Officer of Sealol (among other things, the free world's principal supplier of stern-tube seals for submarines), becoming Vice President and General Manager in 1965.

When Sealol was acquired by EG&G in 1969, Robbins moved up to the parent Company's Massachusetts headquarters to help CEO Bernard O'Keefe oversee various EG&G ventures.

The Sealol experience had given Robbins an appetite for hands-on management at a manufacturing company, and his new role at EG&G did not seem to point in that direction, nor did it offer him a career path equal to his ambitions, and he began telling friends of his interest in moving. The network included Lybrand's partner Gene Friedman, who knew d'Arbeloff and, more to the point, knew that Teradyne was looking for a senior financial manager. At Friedman's suggestion, d'Arbeloff called Robbins to arrange a meeting.

Robbins is, in the opinion of most of his peers, one of the most talented and creative financial thinkers of his era, but he artfully hides his intelligence behind a "country boy from Oklahoma" facade. Many an adversary has innocently walked into the trap, spouting some shallow financial argument that Robbins would then proceed to eviscerate. Tall, rangy, and laconic in the Gary Cooper manner, Owen Robbins wears Pauls Valley on his sleeve; it is part of his weaponry.

Of course, the folks at Lybrand were not fooled; they had in fact tried to lure him back, with a partnership in view, and this led to a brief face-off with d'Arbeloff, who had decided that Robbins was just what Teradyne needed. The decision, of course, was Robbins's, and he chose to join Teradyne.

Why Teradyne? For one thing, his interest in operations could not be satisfied at the accounting firm. "Also," he says, "I had developed a theory about using finance as a major tool in managing a company, and Teradyne would give me an interesting canvas to paint on."

And paint he did, brilliantly, for the next 30 years.

PART THREE

The Turbulent Seventies

11

• • •

Teradyne Goes Public...

With the solid performance of 1969 behind it, Teradyne was ready for the public market. Prospective underwriters made pilgrimages to Boston to present their credentials to the founders and to Gene Lyne, who as Treasurer and chief financial officer was overseeing the process. One investment banker, noted for its aggressiveness, proposed marketing the stock at 100 times 1969's per share earnings of about 50 cents, for an IPO price of $50 a share. DeWolf and d'Arbeloff thought this a reach, even in the frothy market of the day, and finally settled on the relatively conservative and prestigious Lehman Brothers to shepherd Teradyne into the public arena, at 50 times earnings. A total of 425,000 shares would be sold to the public, at $25 a share. Since a number of existing stockholders were joining the Company as sellers, Teradyne's take would be slightly more than $6 million.

At the time of the offering, the stock market was gripped by a mania for high-technology stocks, and it was reasonable to expect Teradyne's shares to soar as soon as they traded publicly. Thus employees and "friends and family" pressed the Company for the opportunity to buy shares at the offering price.

A formula was worked out by the founders and Gene Lyne, allowing each employee to buy a number of shares based on his or her salary. DeWolf and d'Arbeloff went even further, giving each of the current employees five shares of stock out of their pockets.

Teradyne's new role as a public company brought with it new responsibilities for disclosure and communications, and to handle these and other assignments I was shifted from the sales promotion department to an eighth-floor office next to the offices of the two founders, to whom I now reported. Ed Sutherland, who in 1969 had followed me to Teradyne from General Radio, became the Company's new Advertising Manager.

In March, Lehman Brothers escorted the Company's management on the customary "road show" to market the Company to the investment community, a pro forma undertaking, since the issue was sure to be oversubscribed. The Company's first-quarter results did nothing to dampen the enthusiasm. Sales were $4.5 million, double those of the year-earlier quarter, and earnings of 15 cents a share were up almost fourfold. All systems were "go" on April 4, when the first shares were traded on the over-the-counter market at $25. Before the day was out, the stock was trading at $34, and jubilant employees were calculating their windfall.

But two factors were conspiring to rain on Teradyne's parade. First, the "go-go" days of the stock market were just about over. In fact, only one other hot new issue[1] made it to market before the window slammed shut. Second, Teradyne's bookings slowed somewhat in the second quarter, then disappeared in the third. The Company's backlog would enable it to report another growth year in 1970, but the dearth of orders, along with cancellations, all but guaranteed a down year in 1971.

[1]Toys 'R Us

Teradyne was far from alone in its misery. The semiconductor industry and its equipment suppliers were being rudely introduced to the extreme cyclicality that would plague them for the next two decades. The nature of the problem was easy enough to understand: High demand for semiconductors would induce chip makers to base their equipment purchases on overoptimistic sales projections. Then, when these projections failed to materialize, they didn't simply slow down their spending; they eliminated it, even canceling orders on the books. Easy to understand, hard to defend against. The Company would learn to ride the roller coaster better than most, but the cyclical downturns were never painless.

When business turns, Wall Street senses that fact instantly if not sooner, and in a matter of weeks Teradyne stock was trading below its $25 offering price. This produced a great deal of hand-wringing at Essex Street, and the founders, feeling somehow to blame for encouraging employees to buy the stock at $25, were moved to consider a buyback in order to make the employees whole. Gene Lyne, asked to ponder the legality, thought the idea probably illegal and definitely unwise, and nothing further developed along these lines.

As bookings evaporated, the stock price dropped sharply, hitting 14 in early July, then bouncing to 22 in early October, when the announcement of the Company's first layoff — 15 percent of the work force, or just over 100 employees — sent the stock tumbling to $10.50 in November. In mid-October the stock began trading on the American Stock Exchange, under the symbol TER. In the same month, DeWolf updated stockholders on the situation, concluding with what turned out to be a prescient observation:

> "Considering the nature of our business, I expect that whenever the upturn does come, it will come suddenly rather than gradually, and the boom will again drown us in opportunity."

There were solid grounds for Nick's optimism. The first SLOT Machine had been shipped, and it appeared to be a worthy successor to the J259. The new test system for linear integrated circuits had become the leader in a small but promising market. Teradyne Dynamic Systems (TDS) had just shipped its first pulse parametric (i.e., dynamic) test system, and its compatibility with the large J259 base gave it a head start on the competition. Testers for discrete semiconductors (transistors and diodes) still sold well. In Chicago, Teradyne Applied Systems (TAS) had developed three J259 derivatives: one for testing motor controllers, one for laser-trimming resistors, and one for testing circuit boards. In Lowell, Teradyne Components had established itself as a dependable supplier of backplanes to military contractors. With the stock offering behind it, the balance sheet was sound. Thus, in August 1970, with the downturn at its worst, Lehman Brothers' analyst Tony Langham was able to tell his Company's clients:

> "Since its inception, the Company has proven its ability to innovate technologically, and to market its products successfully and profitably. Few companies have compiled a record equal to Teradyne's, and few operate in a total market environment which offers as much potential for sustained high growth. We consider management to be exceptionally competent and are confident that it will continue to drive

> Teradyne towards a position of leadership within the over-all electronics industry. We consider the stock an attractive investment."

The bookings drought was to last from mid-1970 through mid-1971. Given 1970's strong first half, this meant that 1970 would look reasonably good in the record books. Sales of $18.1 million were 18 percent higher than in 1969 and, although margins were squeezed, the Company was still able to report net income of more than a million dollars. In 1971, on the other hand, the recession would stand revealed in all its starkness. Red ink flowed for the first time since 1962, and sales dropped by 28 percent, to about $13 million. The boom was just around the corner and was about to drown Teradyne in opportunity, as DeWolf had predicted, but by then Teradyne would have a new leader.

Teradyne under both DeWolf and d'Arbeloff always believed that the best defense in a recession was a strong offense, and the Company sought to use the dry spells to develop products that could be exploited in the next upturn. Thus, in 1971, red ink or no red ink, Teradyne worked feverishly on a new MOS test system, the Company's first home-grown board test system, a new backplane test system, and a new computer-operated transistor tester. All these systems were controlled by Teradyne's M365 computer, making the Company one of the larger computer makers of that era.

By 1971 the Company logo, the word "Teradyne" balanced atop a triangle, was looking dated, and a replacement was in order. Designer Dave Lizotte, working for Boston's Gunn Associates, had been designing Teradyne's brochures for several years, and he was asked to create the new look so desperately needed. His suggestion: a simple, elegant rendering of the word "TERADYNE" in shadowed block letters. This was submitted to DeWolf and d'Arbeloff, who viewed the matter as an annoying distraction from the real problems of the day but who okayed the change. The shadow-letter logo would prove wonderfully workable and surprisingly durable, identifying Teradyne distinctively for the next 30 years.

1971 began with no noticeable improvement in business. Bookings bumped along bottom at $2.5 to 3 million a quarter — obviously not enough to support even a slimmed-down Teradyne. The Company posted a net loss of $460,000 in the first quarter, but the stock had begun to move higher in anticipation of a recovery and was trading at $17 by the end of the first quarter. And there it stood, more or less, until a memorable presentation by the Company at the New York Society of Securities Analysts on July 14.

Nick DeWolf's products were always safe, dependable, predictable. A Company of substance had been built on Nick's ultra-conservative engineering. Yet, when standing before an audience of securities analysts as President of Teradyne, DeWolf was not going to be safe and predictable. He would say exactly what was on his mind, because his honesty overwhelmed his diplomacy. What was on his mind, that Tuesday morning, was a prediction that the semiconductor test business was about to turn up — not because of any conventional market research, but because, as he declared:

> "The people who are sending us semiconductors are shipping junk."

At the time, the audience of securities analysts paid the comment only passing notice. DeWolf was simply observing, somewhat colorfully, that semiconductors were being inadequately tested, and that once that realization "hit the fan," tester sales would pick up. In that context, the statement was unremarkable.

But reporters never let context get in the way of a good story lead, and so, in the following Monday's *Electronic News*, there was DeWolf's photo on page 1, over a story that opened with the "shipping junk" quote.

The battleship Massachusetts had taken a hit, and it was a time for damage control. D'Arbeloff hastily telephoned executives at National Semiconductor, Intel, Motorola, and other customers to assure them that it was not *their* products DeWolf was calling junk. To their credit, the customers were generally sympathetic to Teradyne's predicament (most of them had probably suffered similar embarrassments), and the crisis ended without great damage to Teradyne. It is true that the stock dropped from $17 to $13.25 in the two weeks following publication of the story, but this was probably due more to the fact that DeWolf and d'Arbeloff used the meeting to telegraph further quarterly losses for the balance of the year.

12

• • •

… AND D'ARBELOFF TAKES THE HELM

Although the "junk" quote had little or no effect on Teradyne's business, it did sting the Company's sales force for a time, and the effect on DeWolf must have been traumatic. As President of a private company, he never had to worry about such nonsense. Was he now duty-bound to watch his every word, lest he be misinterpreted? Would the outside world that now owned Teradyne let Nick be Nick? The frustration that had been simmering for many months now came to a boil. Three weeks after the securities analysts' meeting in New York, Nick DeWolf resigned the Presidency and was succeeded by Alex d'Arbeloff. It is tempting to connect the two events, but the reality is that DeWolf had been considering giving up the presidency for some time, possibly since the discussions following the episode of d'Arbeloff's bleeding ulcer, back in 1968.

Also, perhaps because of d'Arbeloff's well deserved reputation as a master strategist, some were inclined to see his accession to the presidency as a palace coup, but this reading is also provably false, on the evidence of the testimony of both principals.

"Nick was no longer willing to put the time and energy into the job, and this was a problem we both recognized," recalls d'Arbeloff.

"That's absolutely correct," agrees DeWolf. "I was growing bored. My heart just wasn't in the monster machines that were coming, like the one Salvador was working on. And so I plotted my own departure, and Alex and I were very up-front about it."

And so, on Wednesday afternoon , August 4 1971, at a special Board meeting at the Company's Essex Street headquarters, Alex d'Arbeloff became Teradyne's President and Chief Executive Officer, while DeWolf was made Chairman of the Board and given the additional title of "inventor at large." Attending the Board meeting were four of the Company's six directors: DeWolf, d'Arbeloff, Bill Elfers of Greylock, and ARD's John Shane. Director Henry Watts was vacationing in Scotland at the time, and Gene Lyne was on Cape Cod, where he was apprised of the situation by telephone the day before.

"I am resigning the Presidency," DeWolf told me at the time, "because I am an honest man, and the truth is that for some time I haven't filled the role the world expects of company presidents." He had acquiesced as his partner became de facto manager of the Company's affairs. Now his overriding concern was that the changing of the guard be presented to the public, not as a palace coup, but as a move initiated by him. A further concern, especially to the outside Directors, was the effect of the change on the Company's stock price. In the minds of many, after all, Teradyne *was* DeWolf, and the prospect of his loss could be damaging. Thus, once the Board had acted, attention turned to the wording of the announcement. With the second quarter's earnings release imminent, one Board member suggested combining the two announcements, but others correctly saw that this would connote a linkage (the numbers were not good) between the two. Common sense prevailed, and the following release was sent to the Dow Jones News Service and Reuters after the market's close on August 4:

> DEWOLF AND D'ARBELOFF,
> COFOUNDERS OF TERADYNE, INC.,
> ELECTED CHAIRMAN AND PRESIDENT
>
> BOSTON, MASS. — Teradyne, Inc., Boston-based manufacturer of electronic test equipment, announced today the election of Nicholas DeWolf as Chairman of the Board and Alexander d'Arbeloff as president.
>
> Mr. DeWolf, founder of Teradyne and president since 1960, relinquished the presidency in order to devote a greater proportion of his time to the technical side of Teradyne's business. He characterized his principal future role as that of "inventor-at-large," in which capacity he will explore potential areas of new product development for the Company.
>
> Mr. d'Arbeloff, formerly executive vice president, was Mr. DeWolf's partner in the founding of Teradyne.

Earlier in the day, Teradyne's Boston employees were summoned to the Company cafeteria, where Nick DeWolf broke the news, assuring the work force that he was not "copping out" or fleeing the Company in the midst of recession. On the contrary, he argued convincingly, he was sure that Teradyne's prospects were brightening. Nick's spirits seemed elevated by the session, as if he had just been delivered of a terrible burden.

Between the market's close on Wednesday and the opening on Thursday morning, telephone calls were made to certain large stockholders to assure them that the Company remained in good hands. Opinions on the likely effect were sharply divided, some key employees sure the stock would be devastated, others equally sure that there would be no damage. The alarmists were wrong. The stock, which had closed on Wednesday at $14.50, opened Thursday morning at $14.625, moved steadily higher, and closed at $15.25 — all on a total trading volume of 700 shares!

It would be hard to overstate the anguish that must have lurked just beneath DeWolf's cheerful demeanor, despite the fact that his exile was self-imposed. Teradyne was a creature of his genius. He had invested in it years of incredibly hard work, energy, and enthusiasm. He had fashioned the design principles honed at General Electric into a successful, widely respected enterprise. And now he was an "inventor-at-large" at his own Company. Even worse, he had been "promoted" to the position of Chairman of the Board — a title he always loathed.

In a letter to stockholders dated August 10, DeWolf followed the line of the news release, adding the following gracious tribute to his partner:

> … "I am doubly confident of the future because my successor as president will be Alex d'Arbeloff, who joined me in founding Teradyne 11 years ago and who has worked with me more as a partner than as

> a vice president ever since. With the presidency passing to my alter ego, there is absolute continuity of management at Teradyne..."

The trade paper *Electronic News* sought interviews with both DeWolf and d'Arbeloff. D'Arbeloff dutifully endured a meeting with the EN reporter, but DeWolf, no doubt feeling burned by that paper a few weeks earlier, declined. "I really don't want to make a big deal out of this," he told a colleague. Besides, his attention was turned elsewhere. Long an expert photographer, he now wrote to the Chinese ambassador in Ottawa, requesting a visa for a photo-taking tour of that country.

The downturn hit bottom with the third quarter's sales of $3.08 million and a loss of 14 cents a share. Upon the earnings release, in early November, the stock dropped below $11. The faithful who bought at that point were soon rewarded; four months later, when the recovery was visible to all, the stock was trading at $23.

This recovery would propel Teradyne sales from a 1971 low of $13 million to $56 million only three years later. The Company's new president could hardly have chosen a better time to take the helm. More than luck was at work, however. The Company's new products, most of them designed in the later stages of the DeWolf era, were successful almost without exception, and Teradyne, which had earned a place in the electronics industry's consciousness with the J259, was increasingly seen as a broad-line maker of automatic test systems. The flagship product was still a digital IC tester, the SLOT Machine, and transistor testers still contributed to the revenue stream, but in addition there were now a circuit-board test system (the L100), a backplane test system (the N151), laser trimming systems for film resistors and hybrid circuits, and a highly successful back-plane connector operation. The J263 linear IC tester, marketed skillfully by Product Manager Graham Miller, enjoyed clear sailing in a market that belonged entirely to Teradyne.

The term now used to describe Teradyne's core business was "automatic test equipment," or "ATE." D'Arbeloff disliked the term and resisted it as long as possible before allowing its use in Company literature.

The Company's rapid growth in the early 1970s triggered the addition of more space. The Essex Street plant was full, the nearby Lincoln Street plant (the Albany Building) was filling up, and the west coast operation, whose name had changed from Teradyne Dynamic Systems to Teradyne Digital Systems, was bursting its seams in Chatsworth. Teradyne Applied Systems (TAS), the unit spawned as a Teradyne — Cozzens & Cudahy joint venture, was now a Teradyne Division, focused on the fast-growing market for laser trimming systems. At its helm was Warren Cozzens, a giant bear of a man and a born salesman, with a broad, toothy smile, a hearty handshake, and an ability to see humor in any situation.[1] He was also an instant authority on any subject, and within months of first putting his toe into the laser-trim waters, he was the Grand Old Man of Lasers, lecturing old laser hands in the subtleties of micro-cracking. He pulled it off, not because he was guileful, but because he did his home-work. Some called him "The Silver Fox" because of his mop of gray hair, and the term was always used with admiration.

[1]An example of Cozzens humor, circa 1973, is his "Teradyne Creed": "I believe in Teradyne, the Almighty, maker of electronic test equipment. And in Alex d'Arbeloff, its only begotten son, our Lord, who was conceived by Nick DeWolf, born of M.I.T., suffered under Abraham Bluestone, was classified 'reject' and binned.... I believe in the ten-year warranty, the sanctity of documentation, the sealed module, the forgiveness of customer ignorance, and the master price list everlasting."

Far removed from the Cozzens School of Salesmanship was the craft as practiced and as taught by Jim Prestridge, who was busily building Teradyne's new direct sales force. Prestridge played it all by the book, and under his leadership the new Teradyne sales force was trained in the techniques of selling as set forth by the best business schools, the best textbooks, and the best seminars, run by the best consultants. To Prestridge, the sales process was just that — a process — and the journey from initial contact to closing the order was marked by signposts that were easy to read and follow. The process had worked for Prestridge at Hewlett-Packard, and it had worked at countless other companies, and it would work at Teradyne.

The building of a sales force was not going to be easy. There were representatives to be phased out, new people to be recruited, veterans to be repositioned or dropped. Then there was the legacy issue: Teradyne had been arrogant in the past, and some customers had been bruised. D'Arbeloff's brand of selling had not always been gentle. Bluestone had also left some scars. There were bridges to rebuild, but Prestridge was just the professional the job required.

A fact of life that could not be ignored was the emergence of the San Francisco Bay Area as a major center of semiconductor activity. Could Boston-based Teradyne compete in the back yard of its primary competitor, Fairchild? The west coast representative, ISI, was spreading the word well enough, but a direct presence was needed. Thus Dick Bailey, a successful Teradyne salesman in the Northeast, was dispatched to open a Teradyne support facility on Page Mill Road in Palo Alto, in the shadow of Hewlett-Packard's headquarters. There Bailey coordinated the efforts of the local representative and began to "learn the territory." With him in the new office was Gordon Padwick, an applications engineer hired from Fairchild.

Settling in at Palo Alto, Bailey soon found that Teradyne, far from being disadvantaged in the Bay Area, had a secret weapon: the hordes of managers and engineers who had developed loyalties to Teradyne at east-coast companies like Transitron and Sprague and Sylvania and who were now journeying west to join Fairchild, Intel, and National Semiconductor, where they would quickly be added to Bailey's Rolodex.

It is not clear whether Bailey at the time regarded his west coast assignment as temporary, but in fact he never returned, spending the rest of his career selling for Teradyne in Silicon Valley — and never losing a Boston accent that was so extreme it was almost a caricature.

Another professional was building a new organization in Europe. Dennis O'Connell, after learning the Teradyne ropes running the Company's Philadelphia office, returned to his native London in August, 1971, as head of Teradyne's European operations. Succeeding O'Connell as U.S. sales manager was Tony Martinelli, recruited from Hewlett-Packard by Jim Prestridge.

In Europe, O'Connell inherited a strong team of sales managers, mostly recruited by his predecessor, Tony Wakeling. These included Peter Bette (Germany), Francis Giard and Alain Renaud (France), David Newth (U.K.), and Chris Davison, who would later move to Tokyo to manage Teradyne's Japan operations.

Recognizing the potential of the Japanese market early in the game, Teradyne opened its own sales office in Tokyo, phasing out its relationship with Midoriya. Stu Peters, who had followed Nick DeWolf from Transitron to Teradyne a decade earlier, was sent to Tokyo to head the new sales organization, made up, for the most part, of ex-Midoriya employees. At first glance, it was a strange posting. There was nothing in Peters's background to suggest that he would be comfortable in a setting so different from his native Massachusetts — or that his Japanese subordinates would be comfortable with him. As it turned out, the appointment of Peters was a brilliant stroke, and years later, when he was transferred to London, there wasn't a dry eye at his farewell party in Tokyo.

The Board of Directors was expanded to seven with the 1973 addition of Franklin "Pitch" Johnson, a California venture capitalist, and Dick Testa, Teradyne's outside counsel and Secretary. (Director Gene Lyne had by this time left the Company.)

The adrenaline flowed at Teradyne in the early 1970s. The Company was now large enough to offer its employees a degree of security, and it was growing fast enough to provide ample excitement, and yet the small-company spirit and informality nurtured by DeWolf and d'Arbeloff in the 1960s was still in the air. Nothing illustrated the point better than "The Teradyne Follies," an employee-produced entertainment introduced at the Company's 1972 Christmas party. The Follies featured a chorus line of secretaries called The Terabelles, a "jug band" made up of a half dozen employee-instrumentalists, and, most memorably, a series of blackout skits lampooning various aspects of the Teradyne scene. In one skit, for instance, Dick Lucey, playing Teradyne President "Alex Telemoff," reacted to a customer's request for a longer system power cord by insisting that the customer be told to move the wall closer to the system. This send-up of d'Arbeloff's "Teradyne arrogance" became part of the Teradyne lore for decades. Other skits took aim at the tendency of certain sales people to overpromise, the Company's 10-year warranty, and, in the most memorable of all the skits, the Essex Street plant's shaky elevators. In "The Elevator," a visitor, visibly vibrating while riding one of these elevators, asks his Teradyne host why the Company didn't fix them. "Just wait," answers the Teradyner. The elevator then stops, and an attractive, shapely young lady gets on. As the elevator resumes its jiggling ascent, its passengers also jiggle, to the great merriment of the audience. Perfectly cast in the starring role was Personnel Manager Arlene Harrington, whose good looks were exceeded only by her good humor. The audience, made up of employees and their families, loved every minute of it. These were Teradyne's salad days.

Although camaraderie was abundant at the Essex and Lincoln Street plants, there was little socializing after work, for reasons mostly geographic. Some of Teradyne's Boston employees commuted from south of the City, some from the north, some from the west. Some drove in and out daily, while most City residents used mass transportation. There would occasionally be successful attempts to organize basketball or hockey teams, but the vast majority of Teradyne employees saw little of each other outside of working hours.

Another characteristic that set Teradyne apart from suburban companies was its racial mix. Minorities generally represented 15 to 20 percent of the Boston work force, a percentage several times that of the typical Route 128 company. This was a natural consequence of Teradyne's inner city location, of course, and the Company benefited greatly from its position as virtually the only high-tech employer downtown. The pool of unskilled but willing workers was vast, and Teradyne could afford to be highly selective, hiring the best of the hundreds of applicants who responded to every advertisement. The pay and benefits were significantly better than what most other downtown employers were offering, and the subway system, with its downtown stations a short walk away, furnished the transportation.

Most of the minorities worked in production, and those with leadership skills soon became foremen and managers. Social engineers, had they taken the trouble to look, would have found in Teradyne's Boston operations a microcosm of their ideal multiracial world, with people working effectively together without as little regard for skin color as DeWolf and d'Arbeloff had for titles or perks.

Teradyne's role as a major employer of minorities did not totally escape notice. On one occasion a reporter from a Boston television channel asked to visit the plant and interview management. Arriving with a cameraman at his side, the reporter asked a few opening questions, then came to the point of his visit.

"Tell me about your racial strife," he asked.

"There is none," was the honest reply, whereupon the reporter signaled to his cameraman that the visit was over, and left. No strife, no story.

In late 1970, as bookings plummeted, d'Arbeloff began to worry that the semiconductor test business might not be large enough to support the two leading participants, Teradyne and Fairchild Test Systems. The same thought had occurred to Fairchild, and the two competitors began discussing the possibility of merger. These talks reached an impasse in January, 1971, but were restarted in May, when d'Arbeloff flew to California to attend the first Semicon trade show. Talks intensified over the next two months, involving senior executives from both companies, but they eventually led nowhere. D'Arbeloff saw in Teradyne a company on its way up, and his market share data showed Teradyne's sales to have climbed from 90 percent of Fairchild's in 1968 to 174 percent in the first half of 1971. Fairchild, of course, regarded its slide as an aberration soon to be corrected. By late 1971, it was clear that a recovery had begun, and the talks lost any urgency they might have had.

The recovery was real. Shipments of $6.4 million in the fourth quarter of 1971 jumped to $7.5 million and $9.1 million in the next two quarters. The stock market, having been jolted by the 1971 downturn, remained skeptical, however, and Teradyne' stock price languished in the mid-teens until July, when it soared to the high 20s, exceeding its public offering price for the first time in three years, before falling back to the low teens in December. Such volatility would be the rule, not the exception, throughout the Company's history.

The SLOT Machine picked up where the J259 had left off, and production of the new system and the J263 linear IC test system was proceeding apace at Essex Street, under the direction of Production Manager Fred Macdonald, a capable, soft-spoken young Princeton graduate who had joined the Company a few years earlier. To gain entry to the fast-growing market for MOS test systems,[1] the Company developed the J277 MOS Test System. The ambitious, complex system, conceived as a possible gateway into the LSI test market, presented one production hurdle after another, leaving d'Arbeloff exasperated.

"It's like being back in Brazil," he told Prestridge one day, "where you know that something always goes wrong, and the only suspense lies in how it goes wrong. You send a guy across the street for a package of cigarettes, and you know he won't

[1]It was then the practice among ATE analysts to divide the market into bipolar and MOS market segments.

come back. You send him ten times and the result is always the same, only he's got a new story every time."

Organizationally, the IC test operation was far from ideal. Jack Salvador, now presiding over an organization responsible for the lion's share of ATE revenue, remained ensconced in the San Fernando Valley, although some of his most important systems, such as the linear test products marketed under Graham Miller, were produced in Boston. Salvador had become a major player at Teradyne but had never worked in Boston and had never developed a close working relationship with d'Arbeloff. D'Arbeloff thus found it easy to second-guess Salvador's decisions, and Salvador, 3000 miles from the Company's power center, was quick to sense real or imagined slights.

Although Nick DeWolf would remain Teradyne's Board Chairman through 1975, his involvement in the Company's affairs dwindled steadily. In January, 1972, he left for an extended tour of the Far East, promising to return in time for the Company's annual stockholders' meeting. After his return (with hundreds of magnificent photographs taken in Bali, Thailand, etc.) his interest in Teradyne waned almost to the vanishing point. Also, as the character of Boston's Beacon Hill changed in the early 1970s, DeWolf began searching for a new home. In the Company's 1976 Annual Report, the Chairman is listed as "Nicholas DeWolf, Aspen, Colorado."

The vicious downturn in semiconductor test sales that hit the Company in 1970 convinced d'Arbeloff that diversification had to be an essential part of Teradyne's strategy. The process had begun with Teradyne Components and with a number of testers based on J259 technology, but more was required, and in late 1971 Teradyne's new president began surveying the possibilities. The automotive business seemed a promising potential market for ATE, and a small California-based maker of ignition and exhaust analyzers was considered. A credit-card verification service was also discussed. An idea for a highly reliable, inexpensive line printer was promoted by Teradyne mechanical designer Walter Kern. None of these held any appeal for d'Arbeloff, but he kept looking.

The semiconductor test business, meanwhile, had turned from difficult to nearly impossible. A major semiconductor maker (and an important Teradyne customer) threatened to put existing tester orders on hold unless Teradyne agreed to buy more of its integrated circuits. Another IC maker decided to quit the business, dumping its Teradyne test systems on the used-equipment market. Long-time competitor Fairchild Test Systems, after having been staggered by the J259, was back in the game with its own computer-operated testers, sold at bargain-basement prices (subsidized, Teradyne suspected, by its deep-pockets parent).

As Teradyne's sales quadrupled in the first three years of his presidency, d'Arbeloff was riding high, but there were inevitable strains, both on the organization and on its leader. The management pyramid was flat beyond reason, with more than 20 people reporting directly to the President. The Company's only vice president in 1972 was Jim Prestridge, who was given the title in November, 1971.

Titles were a sore spot with d'Arbeloff, and in naming Prestridge a vice president he took pains to explain that the title was given, not to reward performance, but only to enable Prestridge to carry out his duties more effectively. The outside world, it seemed, placed some store in titles, and the Company would therefore reluctantly confer one on its sales manager so that he could deal effectively with customers. The same attitude covered organization charts (there were none) and business cards, all of which were without titles. The rationale: A growing high-tech company needed to be flexible, and titles and organization charts, like unions, tended to make a company rigid and excessively dependent on hierarchy.

D'Arbeloff also held that employees should not be distracted by inconsequential matters such as titles or other perks. The relative contributions of various members of the work force should be recognized by salary and nothing else. Such egalitarianism was not entirely of d'Arbeloff's making; it was deeply ingrained in the DeWolf psyche. The egalitarian culture was also modeled after that practiced at Hewlett-Packard, a high-tech icon, and at Intel, which was fast becoming one. Teradyne was thus patterning itself, not on the class-bound Brahmin institutions of its home city, but on the open-collar companies of Silicon Valley.

Teradyne's competitive landscape was becoming more interesting as the 1970s unfolded. The constant in the mix was always Fairchild, but there were also a number of smaller companies, such as Redcor, Datatron, Macrodata, and Lorlin, all striving for their places in the sun. And the sleeping giants — Hewlett-Packard, General Radio, and Tektronix, were awakening to the possibilities in semiconductor testing. Two H-P engineers were dispatched to a Teradyne memory-test seminar held in Woodland Hills, California, in February, 1971, with the express assignment of intelligence-gathering. (Turnabout was only fair play, and a concerted, nearly successful effort was mounted to recruit one of these engineers and his entire design team to Teradyne.) General Radio was marketing an automatic board test system in competition with Teradyne's new L100. And Tektronix was readying an assault on the dynamic IC test market.

The success of the Woodland Hills memory-test seminar prompted the Company to stage an east-coast version, at New York City's Plaza Hotel, and this meeting produced a hilarious vignette. About 100 attendees endured a morning listening to Jack Salvador's lecture on the art of testing semiconductor memories, a session that generated more drowsiness than enlightenment. A full hot lunch only worsened the problem. Then, in mid-afternoon, with the audience thoroughly numbed, the seminar organizer, standing against a glass door at the rear of the room, sensed someone standing on the other side of the glass, peering in. It was Red Skelton.

"Come on in," said the Teradyner.

"What's going on?" asked Skelton.

"It's a memory test seminar."

"You mean, like test your memory?"

"Not exactly, but come in anyway."

Skelton needed no further encouragement. He swung open the door and strode majestically up the center aisle of the room, the audience agape. Once at the front of the room, he held up his arms and said: "I just want to tell you all that Mamie Schmidlapp has just pledged two hundred dollars." With that, he walked out, to the applause of the attendees, who were convinced that Teradyne was indeed one clever company to have staged such a star turn.

In January, 1972, American Research & Development, one of Teradyne's oldest and largest shareholders, announced its intention to merge into Textron, Inc., a large conglomerate. ARD's 400,000 shares (plus an option on an additional 88,000 shares) amounted to more than 15 percent of Teradyne's outstanding stock, and speculation was rife as to its disposition after the merger. Some of Teradyne's larger institutional stockholders feared that Teradyne might conceivably be absorbed by the larger company, but in the event, the merger proceeded without any noticeable effect. Textron held its Teradyne shares for several years and then sold them at a cyclical low point.

Only a few weeks after ARD's announcement, there was another bit of unsettling news: Milt Collins, the father of Teradyne software, and Abe Bluestone, one of the Company's leading sales executives, disclosed their intention to leave Teradyne to start a new company to apply the considerable advantages of the Teradyne M365 computer to business applications, including inventory management. The two had floated the idea of packaging their company as a Teradyne-funded venture, but d'Arbeloff considered the business plan flawed and passed. It was a bittersweet pill: The two had contributed enormously to Teradyne's past success, but it was time to move on. Bluestone had been replaced as sales manager by Jim Prestridge. As for Collins, the stimulation of working with DeWolf was now gone, and the same driving ambition that had taken him from Transitron to Teradyne was now calling him elsewhere. In this case, "elsewhere" was not that far away, for the new company, called Minico Systems, was soon to sign Teradyne as its first, largest, and, as it turned out, only customer. Soon M365 computers equipped with Minico software were on line managing Teradyne's inventory. What's more, Minico could even boast a d'Arbeloff on its payroll: Brit, Alex's wife, had decided to dust off her engineering degrees and write programs for the new venture.

Alex, who never enjoyed defections, had mixed feelings about the loss of Collins and Bluestone. He was especially close to Bluestone, who had worked so brilliantly to launch Teradyne in Europe, and he knew that Collins was a formidable talent. But in this case his disappointment was softened by the ties that would bind Teradyne and Minico. Besides, there were increasing signs that the recession was ending, capped by the receipt, in early March, of the largest single order ever placed for semiconductor test equipment. The customer was National Semiconductor, and the sixteen systems included both digital and linear IC testers.

In Lowell, meanwhile, Windy Hunter, Tom Neavitt, and Len Johnson were proving to be a winning management team, and Teradyne Components was rapidly outgrowing its quarters. After considering the acquisition of adjacent land in Lowell, the trio decided instead to purchase a plant site in Nashua, New Hampshire.

13

• • •

Growth and Diversification

By mid-1972, it was clear that the semiconductor industry was in a buying mood again. Yet the memories of the recession were not easily forgotten, and d'Arbeloff remained convinced that salvation lay in diversification. He was especially intrigued by the possibility of applying Teradyne automated test techniques in the telecommunications industry. "I read an article that said AT&T made something like 700 tests every day on its telephone lines," he recalls, "and it occurred to me that our testers routinely made thousands of tests on a semiconductor in seconds, so we ought to be able to bring something to the party." He initially flirted with the idea of entering telecommunications through acquisition. One prospect was a small New Hampshire company called Northeast Electronics. But the owners of Northeast demanded what d'Arbeloff considered an outrageous price for their company, and d'Arbeloff thus decided to build rather than buy a phone-line test company.

He had near at hand just the engineers he needed to lay the foundation for the new business. Russell Ashdown, who had joined Teradyne in 1968 by way of Mark Instruments, had spent the next few years directing a Teradyne development team (the remnants of Mark) in the U.K., then accepted d'Arbeloff's invitation to come to Boston on a two-year assignment to integrate Mark's analog test technology into Teradyne's IC testers. Chuck Pearson of Teradyne Applied Systems had also moved to Boston in 1971 to join the Central Engineering staff. D'Arbeloff sensed that both were eager for a new challenge, and at his urging they drew up a business plan around a new type of test system that would be sold, not as production-line capital equipment, but as a tool that telephone companies might use to improve their operations.

The pair presented their plan to d'Arbeloff in September of 1972 and found him receptive. Ashdown thereupon flew back to the U.K. to wind up his affairs there (his Mark development team was still on the Teradyne payroll), and while in England he sketched out a design for an automated test system for telephone central offices. Back in the U.S. in early 1973, he and Pearson set up shop in Chicago (Pearson's personal preference). There Ashdown convinced some of the Mark engineers to join the new venture, which was coyly called Teradyne Central — a name that could connote its midwest location or, to those in the know, the telephone "central" office (as in "hello, central").

The architecture of Ashdown's line test system was inspired. The test hardware was distributed between two primary test sites: the COLT (Central Office Line Tester) and the SAC (Service Area Computer). A number of COLTs were equipped with enough test technology and intelligence to run nightly tests on groups of phone lines, and the COLTs would then upload these test results to the SAC, which would further analyze the data and generate English language summaries. The whole affair was orchestrated by advanced multi-tasking software contributed by Teradyne's Boston software engineers.

The Teradyne system would be able to "routine" an entire service area each night, while most telephone users were asleep, and then print out a summary report the next morning identifying failing or suspect lines, with the lines classified by fault category. Nothing like this had ever been achieved. In fact the telephone companies' primary test instrument at the time was a "test desk," where an operator would respond to a customer's complaint by manually executing a test, watching a meter movement, and divining (based on experience) the cause of the problem.

To anyone comparing the two test techniques, the Teradyne advantage was overwhelming — so overwhelming, in fact, that it seemed just a matter of time before the new system would conquer the telephone world. Pearson gave the system a clever trade name: 4TEL. By far the best trade name ever to be coined by Teradyne, 4TEL in time became the term most people used to define Teradyne's line-test business.

No matter how wonderful the technology, few if any products are able to sell themselves, and although Ashdown and Pearson were understandably excited about the product, they soon ran into several unpleasant realities. First, the telephone community, like the automobile community and most old, established industries, was a club, membership in which could be gained only by years of personal associations. Teradyne was not a member of this club.

Second, all of Teradyne's marketing and sales experiences were based on the proposition that customers absolutely had to have what Teradyne was selling. To a semiconductor maker, a test system was not an option or a luxury; it was an essential part of the production line. No tester, no production. The typical telephone company, on the other hand, had been getting along well enough with its antiquated manual test desks and did not feel compelled to change.

Third, Teradyne instinctively began selling 4TEL on the basis of cost savings, while the telephone company was more susceptible to a sales spiel that promised to increase revenues.

There was another obstacle: In order to prove that the 4TEL system was viable, Ashdown and Pearson needed telephone lines, in quantity. When designing a transistor tester it is always easy enough to find a few hundred transistors, but where does one find telephone lines? Fortunately, Pearson had a contact at GTE's Wisconsin company who was willing to provide the trial lines, first at a site near Pearson's summer home on Lake Michigan, later at a full-scale trial at Sun Prairie, Wisconsin, a suburb of Madison.

The trial was a resounding success, GTE Wisconsin's Chief Engineer Clint Davis became a strong sponsor of 4TEL, and word started spreading that 4TEL could revolutionize a telephone company's maintenance and dramatically improve the quality of its service. Other GTE companies began asking for trials, and soon Teradyne was knocking at the door of the telephone club. United Telecommunications (later Sprint) asked for a trial at its Rocky Mount (NC) service center, and at United's Kansas City headquarters senior management began planning for system-wide deployment.

The prize, of course, was AT&T, then a single company representing more than 80 percent of all the telephone lines in North America. Without AT&T, Teradyne could (and eventually did) build a sizable business. With AT&T, Teradyne could be a colossus. So it was with great excitement that a 4TEL system was shipped to AT&T's service center in Hyannis, on Cape Cod, for an extended trial. By all accounts, the system acquitted itself handsomely — handsomely enough to encourage AT&T to accelerate development of its own automated subscriber-line test system, called MLT (for mechanized loop tester).

In time, the MLT system would sweep much of the AT&T network, although it was clearly less elegant than 4TEL, thanks largely to a feature Teradyne couldn't match: compatibility with AT&T's record-keeping system, LMOS. Without access to this proprietary data base, Teradyne was left out in the cold. There were fleeting thoughts of forcing the issue through legal pressure, but these were quickly shelved in view of the volume of business Teradyne was doing with AT&T's manufacturing arm, Western Electric. So Teradyne licked its wounds and sought and found greener pastures elsewhere.

In mid-1972, Teradyne's Production Department was organized into four groups, including one core group and three groups assigned to various product lines. All reported to Production Manager Fred Macdonald. At about the same time, Alex d'Arbeloff was interviewed at length on his perspectives on management and on the relative advantages of a formal vs an informal organization. The interview was never published, but a transcript exists, and an excerpt is well worth including here for the insights it provides into d'Arbeloff's philosophy of management.

Q. If you were to describe the kind of people Teradyne needs, the kind that characterize Teradyne, what would you look for?

A. First of all, you're looking for people who really want to do work. That doesn't mean that you don't need managers, but usually I look for people who want to do something. That's one characteristic. Another is that he thinks through what he's doing rather than just doing it because conventional wisdom says to do it. I have this problem with _____ in personnel, who wants to ignore capable people because they aren't EEs. If we find someone who has something to offer, let's not throw him away because he's not an EE. That's ridiculous. If we threw out all the non-EEs here, the Company would be a lot poorer.

Q. If you concentrate on getting people who are interested in doing things, then you're leaving it to osmosis to develop managers. Do you think that takes care of itself?

A. No, I don't think it takes care of itself. What you want to do is pick out the managers, the fellows that really have skill.

Q. What kind of skill?

A. First of all, to be a good manager, you've got to take your personality out of most of your decisions, and a lot of people can't do that. Some outstanding people are not good managers because they want to think creatively and they're not willing to listen dispassionately to a lot of the problems people have.

Q. That brings me to ask whether the characteristics you look for in people would lead you to select people who would not make good managers.

A. Not necessarily. It would be pretty hard to find a guy who can manage better than Fred Macdonald. We can certainly use outstanding managers who also know our business. We can use as many as we can get. There's a line — a very fine line — between a manager and an entrepreneur. An entrepreneur can not only manage but can also initiate. Then there's the difference between a manager and a leader. A leader can inspire people to do things, while a manager is there to supervise them. I prefer leaders and entrepreneurs to managers, but we still need managers. What I don't want is for every able person in the Company thinking he's got to be a manager in order to make out. If the goal is to become a manager, then you've got a problem.

Q. Then you must have some other way of letting people know they're doing well.

A. There are two things: One is being able to select the job you want instead of having the company select it for you. The other is money. But the thing that motivates people overwhelmingly above anything else is interest in what they're doing. If you can turn them on about what they're doing, that gives them more rewards than anything else. Then you can't ignore the fact that people don't want to be taken advantage of, from the point of view of comparing themselves to the world outside the company and to what they perceive is the pecking order within the company.

Q. So pay is important.

A. If you don't pay enough, it's a negative, but if you do, it still doesn't necessarily make the guy happy. You can pay a guy a lot and he'll still be looking for something else if he's not turned on about what he's doing.

Q. We have a pretty informal organization. Do you think we can stay as loose as we are as we grow into a medium-sized company? As we get into the 30 to 40 million dollar range, will we have to become more structured?

A. We're going to do everything we can to keep the present style. I'm positive we can keep it through the next level of growth.

Q. Is divisionalization one way to accomplish that?

A. Yes. For example, we're now splitting the production department into four parts.

Q. Will these groups relate to the marketing and engineering divisions?

A. Yes. One group will have network and backplane, another will have linear, TAS, and relay test, the third digital systems and discrete, and the fourth will be a central group. In addition to that we're going to split up the test department. Each product group will have its own test department, except for the relay group.

Q. Will this increase the number of people reporting to you?

A. No, because these production groups will all report to Fred Macdonald.

Q. So as all these groups and divisions grow, you hope they will adopt the same kind of loose structure the Company has had. You have often said that the benefits of informality outweigh the disadvantages. Could you recap some of the disadvantages?

A. When you have structure, supposedly you know where everything is done, and everyone knows whom he reports to. Another advantage is that an outsider coming into the company or a stranger studying the company has a way in which he can understand at least how the management of the company intends the organization to run. That's what structure provides. The bad news is that the organization really isn't run that way, because an informal organization exists that reflects the talents and interests of the people involved. The other problem with a formal organization is that it essentially uses management as a reward, and I think that's a disastrous way to run anything. Also, in a formally structured organization the work is always being done by junior people, because the managers are always in meetings.

Q. You're opposed to meetings?

A. I'm opposed to meetings if they're used as a method of making decisions. I'm not opposed to meetings if they're used for information exchange.

Q. It seems to me that if you wanted to keep the most fluid, least formal, least structured organization, the way to do it would be to have everyone in one big group. Divisionalization is itself a structure, isn't it?

A. Well, you're right: Keeping everyone in one big group gives the most flexibility. But I don't think the objective is flexibility. The objective is achievement. And if you want achievement, you need feedback, and the best way to get feedback is to keep people closer to the market, closer to the intrinsic problems of the business, and away from organizational politics. The organization we have is designed to do that.

Q. But you have a company with some 20 groups, each with a de facto manager. If you restructure so that you have, instead of 20 groups, 40, it seems to me you're going to need 40 leaders, or managers.

A. Yes, but as these leaders do their jobs, some will come to supervise more than one group.

Q. You see a natural coalescence over time? Is that what's going on in the present reorganization?

A. I had an argument with Fred on this point. In choosing his managers, he said, "Let's put Chamillard in charge of this, Burke in charge of that, and so on. That's not the way I would do it. I would just announce that this is the way we're thinking about it, and then let the guys choose the way they'd like to go. There are four obvious leaders in the production area — Chamillard, Dias, Cary, and Burke.[1] Why not let them decide? As much as possible I'd like people to think they control their careers instead of us building careers for them.

[1]As usual, his analysis was on the money. All four were still with the Company in 2000, one as CEO, two as vice presidents.

Q. As you said, people do want control over their jobs.

A. The main thing is that people should motivate themselves. All the company can do is create an atmosphere in which this is possible. One thing I forgot to mention before is that one of the real advantages of an informal organization is that it permits creativity, because it makes it easier to change things.

Q. Speaking of creativity, where would you like to see the responsibility for product selection placed in the organization?

A. Again, I don't want to formalize because I think ideas can come from anywhere. What you've got to do is be on the lookout for guys who think outside the box. Pete Sylvan, for instance, thinking about totally new products outside of testing. Ken Skala thinking about retesting in depth. Mann Young thinking about peripherals. None of that is formalized.

Q. I understand there's an informal way of presenting ideas for acceptance or rejection. What about the actual decision to move forward?

A. You have to ask — and this is very important — you have to ask whether the guy is asking to spend his own time on the project, or is he asking for a million dollars to try something out. One of the tremendous advantages that electronics has over, say, mechanics, is that you can try things out very easily. You can empower one or two people to try something out — anything you want to do. And if it's not happening, it's my job to give it a push.

Of the 2.8 million Teradyne shares outstanding at the end of 1972, 43 percent were held by the two founders and ARD, another 16 percent by mutual funds and other financial institutions, with T. Rowe Price heading the list. Early investors in the Company held about 10 percent of the shares, with the balance spread among the employee base. Thus the small "control group" still held enough shares to ward off any would-be predators, and the "outside" ownership was still dispersed. In the Board room, d'Arbeloff could generally count on the support of the Directors, none of whom (except of course DeWolf) had more than a superficial knowledge of the Company's technology and markets. Teradyne was under the firm control of the man one executive reverently called "el Supremo." It was the role d'Arbeloff had been born to play.

The fourth quarter of 1972 brought bookings to a surprisingly high \$10.4 million, and the year-end backlog stood at \$9.4 million, versus a year-earlier figure of \$2.7 million. The stock jumped from \$17 to the mid-20s in January, despite the news of an impending secondary stock offering, with Nick DeWolf among the sellers. The registration was filed on February 26. Lehman Brothers would manage the offering. But it never happened. In March DeWolf, apparently unhappy over the stock price, pulled his shares, and the deal was dead.

The unwinding of the secondary brought a number of players together for intense discussions. In addition to would-be seller DeWolf, there were Teradyne Directors John Shane and Bill Congleton, (ex-ARD officers, principals of the venture capital firm The Palmer Organization, and potential buyers of the DeWolf stock), plus representatives of Lehman Brothers, d'Arbeloff (who remained uncharacteristically passive during the often heated discussions), and Owen Robbins, who was nominally Teradyne's Treasurer but who was fast extending his influence over most of the Company's operations. An abiding interest in the arcana of manufacturing had slowly but surely moved the Company's production resources into

Robbins's orbit, and with an eye to the future he had already hired a possible successor, an accountant named Bill Siddall. It would be 1977 before Robbins would become a vice president, but long before that he had established himself as de facto Chief Operating *and* Financial Officer of the Company, with responsibility for Manufacturing, Accounting, Finance, Personnel, Purchasing — in short, virtually everything except Sales and Engineering.

Robbins also assumed control of the Company's planning processes, bringing order to a function that had been ad hoc at best. Where d'Arbeloff had run the Company with his eye on the P&L statement, Robbins saw, correctly, that any ATE company was ultimately hostage to its ability to forecast bookings and to manage its balance sheet. In the fast-moving ATE business, one never built test systems "for stock"; the inventory exposure would be too great. Indeed, more than one ATE manufacturer had foundered on the shoals of inventory write-downs. Robbins was determined to steer clear of those shoals. Seeing the need for better and more regular oversight, he pieced together the planning processes under which Teradyne would operate for decades to come.

His primary instrument was a quarterly ritual called, somewhat clumsily, the "results/forecast meeting." Its objective was to extract from the Company's assembled division managers a review of the quarter just ending and a forecast of the quarter ahead. The Company's top management would then integrate these reviews and forecasts into a sensible overall near-term plan. Budgets, purchasing, "build plans" — all were integrated by Robbins's results/forecast meeting, which set the tempo for the decentralized Company and allowed it to function smoothly.

Robbins was also responsible for another innovation: the inclusion of the Company's Form 10-K in the Annual Report, starting with the 1973 Report. The Form, which contains a far more detailed summary of a Company's performance than the usual Annual Report, was generally available to stockholders on request, but Teradyne was one of the first companies, if not the first, to include the 10-K in its report — a practice that was applauded by the press[1] and soon copied by other companies.

After the surge in bookings at the end of 1972, a first-quarter slide was expected, but instead the orders set another record, exceeding $10.5 million. The stock price, after its January giddiness, settled back to the mid-teens. The chief culprit was a sick stock market, thanks partly to the daily onslaught of the Watergate hearings. There was widespread discontent at Teradyne over the stock price, which now carried an even heavier burden than usual, owing to a new employees' stock savings plan whose chief asset was an option on 70,000 shares of the Company's stock.

Yet, despite the depressed stock price, the recovery was in full flower, with bipolar IC, discrete semiconductor, and linear test systems all going strong, and Teradyne Components making great progress. A new tester for relays was doing well under product manager Jerry Jodice, and a line of network (or backplane) testers, based on Pete Sylvan's ingenious daisy-chain fixturing, was also thriving under product manager George d'Arbeloff.

The Chicago-based Teradyne Applied Systems had found in its laser-trimming systems a highly marketable product line. The customers were makers of film resistors and hybrid circuits.

[1]Said the *Boston Globe*, on May 7, 1973: "On the basis that stockholders are entitled to as much information as they can get, the Teradyne move is commendable."

In the former case, a laser would remove material from the resistor while the system monitored the increasing resistance. When the resistance neared its target value, the system turned off the laser. The process was fast and much more precise than the sand-blasting systems formerly used to abrade the resistors. In the case of hybrid circuits, the system could monitor some parameter of interest (e.g., circuit gain) while a resistor was trimmed. As the value approached the desired value, the system could direct the laser to make a second, finer cut, further enhancing the system's inherent accuracy and precision. Some years later, a third laser-trimming application would emerge and dwarf the other two: the use of a laser-based system to repair semiconductor memories.

The Company's "back room" housed a new industrial controls group, staffed by Sylvan, Byron Brooks, and two engineers hired from IBM and Gates Radio. The project here involved a counter-timer-controller capable of automating a virtually limitless number of functions. The technology was sound, and the brief but fascinating story of Teradyne's flirtation with industrial controls deserves to be told.

The basic controller was the first Teradyne product to feature imbedded intelligence, initially in the form of an Intel 8008 microprocessor, later upgraded to an 8080. The first units, designated the CS207 and CS217, used Hewlett-Packard magnetic cards to store programs. The smaller CS107, designed to serve as an x-y positioner, used paper tape program storage. Sylvan's idea was elegantly simple: Give the world an industrial controller, built to Teradyne standards of reliability, that would incorporate the latest advances in computer hardware and software. The applications for such a product seemed limitless.

The controller promised to open up entirely new markets to Teradyne. One early unit, the CS210, was sold to Gillette to control inspection of blade alignment in that Company's new TRAC II razor. Another, the dual-processor CS109, was sold to United Shoe Machinery, which used it to stamp shoe eyelets out of brass strips.[1] The CS230, designed to control wire-wrapping in electronic production, found use in Teradyne's operations. But the most impressive controller of all was the CS401, which served as the heart of a pin-straightening machine used with great success by Teradyne Components. The pin straightener was connected to an optical inspection system which scanned a backplane to ensure that all the pins were straight and thus ready for wire-wrapping. Finding an out-of-tolerance contact, the controller would instruct a robotic arm to nudge the wayward connector into alignment. A television screen monitored the whole process, and visitors to the Teradyne Components plant never failed to be impressed by this display of proprietary robotics. The automatic pin straightener, which remained in use long after Teradyne had exited the controls business, proved such a competitive asset that Components General Manager Windy Hunter insisted that none be sold outside the Company.

After a few years of effort, Teradyne's Controls Group was disbanded. The problem, as in the case of the 4TEL System at Teradyne Central, was not the technology but the inability to construct a viable marketing strategy. D'Arbeloff initially considered the Controls unit to be potentially "the most important thing we do," but Owen Robbins and Jim Prestridge, on whom he increasingly relied for counsel, were

[1]Years later, in an effort to eliminate a potential support liability, Teradyne moved to buy back the units shipped to USM — and found the customer unwilling to part with them.

skeptics. So, while the Company stayed the course with 4TEL, it threw in the towel in the case of controllers.

There were thus enough bets on the table (including the most exciting of all, 4TEL) to satisfy d'Arbeloff's hunger for diversification. Managing them all was another matter, especially with an organization in which virtually all matters of substance had to be resolved by the President.

D'Arbeloff's direct involvement in nearly every critical decision would have exhausted a person of lesser energy, but d'Arbeloff, whose near-obsession with tennis helped to keep him in excellent physical condition, stayed on top of it all remarkably well, at least in Boston, where subordinates were within what one Teradyne officer called his magnetic field. That magnetic field did not extend to California, where Jack Salvador's bipolar test division was now accounting for almost half of the Company's ATE business, and despite prolonged phone conversations and countless flights between BOS and LAX, d'Arbeloff's personal influence was attenuated by the 3000-mile separation, a fact he recognized and anguished over.

In mid-1973, Nick DeWolf was officially dropped from the Teradyne payroll. His final project had been an automated color-film processing system, which was obviously outside Teradyne's circle of interest (though very much within DeWolf's circle of competence). Henceforth he would be available to Teradyne, on a limited basis, as a per diem consultant. Thus an illustrious chapter of Teradyne's history — and indeed, of the history of the electronics industry — came to a close, so quietly that few Teradyne employees of the day were aware of it.

The strong recovery of 1973 vindicated d'Arbeloff's drive for product diversification, as robust sales of backplanes, laser trimmers, network testers, and relay testers helped offset disappointing bookings of IC testers. Geographic diversification, another d'Arbeloff goal, also helped. In Europe, Dennis O'Connell's sales team was contributing more than a quarter of the Company's semiconductor test bookings, and in Japan, Stu Peters was proving himself to be an effective leader. Sales in Japan grew at a frustratingly slow rate, but everyone understood that the idea was not to score quick kills but to build long-term relationships. Japan was not yet widely viewed as an emerging electronics superpower, and Teradyne's early direct presence in that market would give the Company a competitive advantage over other U.S. ATE companies.

Sales climbed in every quarter of 1973, from $5.2 million in the first quarter to $9.1 million in the fourth, with profits on the same track. The stock market finally saw what was going on, and Teradyne stock rose from $15 in June to $27 in October, before collapsing to $13 in December, when the mid-East energy crisis caused widespread despair.

The extraordinary growth of the business in the two years following the downturn of 1971 can best be illustrated by a quick look at bookings for the various product lines over this period (numbers in millions of dollars):

	1971	1972	1973
Digital IC test	5.6	9.5	15.4
Discrete semi test	2.8	3.8	8.4
Linear IC test	2.3	6.6	10.3
Laser trim	1.4	2.4	4.7
Board test	0.4	1.2	3.3
Other test	0.2	1.3	3.7
Teradyne Components	1.5	2.7	4.6

In other words, there was growth everywhere one looked, and the awful experience of 1971 was a distant memory. Happy days were here again.

The Company could also take pride in its ability to ramp up shipments in response to its customers demands. Quarterly shipments increased from $3 million to $13 million in the nine quarters since the 1971 low, and close observers could sense that the Company's manufacturing prowess was no less an asset than its engineering capability. Most of the elements of its test systems were produced in-house, and vertical integration was also in evidence at Teradyne Components, where the production floor was filled with home-grown assembly and inspection machinery.

Growth taxed capacity, and with the Essex Street plant now fully utilized, the larger building around the corner on Lincoln Street was being modernized and adapted to Teradyne's needs. In California, Teradyne Digital Systems had outgrown its Chatsworth plant and was investigating other sites in the San Fernando Valley. Teradyne Components opened a small plant in Fullerton, California to serve West Coast customers. This Division's headquarters was still nominally in Lowell, Massachusetts, but most of the expansion was taking place across the border, in Nashua, New Hampshire.

Teradyne Components was not alone in choosing New Hampshire as an alternative to Massachusetts. The high-salaried managers and engineers that staffed Bay State high-tech companies were becoming increasingly unhappy about the tax rates in "Taxachusetts," and soon the Massachusetts high-tech community found itself at war with Michael Dukakis, elected Governor in 1974. The principal outlet for their frustration was the newly formed Massachusetts High Technology Council, whose Board members included Analog Devices' founder and

CEO Ray Stata, Data General's Edson DeCastro, Alex d'Arbeloff, and Alex's brother Dee, now CEO at Millipore.

D'Arbeloff and Dukakis lived in the same Brookline neighborhood, and thus it was natural for Dukakis, in the midst of his 1974 gubernatorial campaign, to ask his neighbor if he might "press the flesh" at Teradyne's Boston plants. D'Arbeloff, busy with a customer, asked one of his managers to escort the candidate through the Essex and Lincoln Street buildings. The tour went badly, with several Teradyne engineers responding to Dukakis's laments about the State's "Great Unmet Needs" by telling him flatly that they didn't care; their taxes were too high.

The warfare between the liberal Dukakis and the generally conservative high-tech community continued throughout his first term, and during that period dozens of Massachusetts manufacturers opened plants in southern New Hampshire. (Today Teradyne's Connection Systems Division is one of the Granite State's largest employers.) Dukakis was defeated for reelection in 1978 by conservative Democrat Edward King, then returned to office in 1982. In 1988 he would run for President as the architect of "The Massachusetts Miracle."

In 1980, the high-tech community, frustrated in its attempts to lower tax rates through the legislative process, helped to fund the efforts of a grass-roots tax-cutting group, Citizens for Limited Taxation, in support of a voter initiative to cap real estate taxes at 2.5 percent of market value. "Proposition 2-1/2" pit the high-tech community directly against most of the legislature as well as the State's powerful teachers' unions. There were tense moments along the way, and the union forces tried by leaflet to turn high-tech workers against their managements. But in this case, the High Technology Council found itself squarely on the side of the angels — and the public. Proposition 2-1/2 won handily, and the Council, which both Alex and Dee d'Arbeloff would chair at various times, was established as an effective political counterweight to the free-spending mentality that had long held sway in Massachusetts.

As he wrestled with the many issues buzzing around Teradyne, as well as the affairs of the High Tech Council, d'Arbeloff also found time to help his alma mater reform its engineering curriculum. An earlier mission to MIT had proved disastrous, when then-Provost Jerome Wiesner turned a deaf ear to d'Arbeloff's complaints about the Institute's curricular shortcomings. "If we don't shape up," he told Wiesner, "the Japanese will take us." Now, a few years later, he was back, as a member of MIT's Committee on the Future of Engineering Education, again warning that the Institute was not adequately preparing its students to work in industry.

While serving on this committee he met a young, dynamic MIT professor named Joe Lassiter. Almost immediately, d'Arbeloff went into his recruiting dance, sure that Lassiter would be a prize catch for Teradyne. It was irrelevant that Lassiter's field was oceanography, not electronics. (His doctoral thesis at MIT was on the Propagation of Water Waves Over Sediment Pockets, a matter of consuming interest to oil drillers.) D'Arbeloff guessed (correctly) that within a year of joining Teradyne Lassiter would know everything worth knowing about ATE.

Lassiter, who had planned to take a sabbatical in 1974, was intrigued enough by d'Arbeloff's sales pitch to spend some time in Teradyne's test department during the first quarter of that year, and he officially joined Teradyne in May. His first major contribution was as co-author of what came to be known as "The Green Book,"[1] a summary of the Company's first principles of management.

[1]More formally, "Teradyne's Basic Strategies for Success in the Test Equipment Business." See Appendix II.

Lassiter brought to Teradyne an ingredient that had been largely missing since DeWolf had left the scene: unbounded, flamboyant enthusiasm. The other members of d'Arbeloff's team — Prestridge, Robbins, etc. — were all uncommonly able, but hardly gregarious. Lassiter lit up the room, and he would not hesitate to gush, "That's fanTAStic!" when so moved. His enthusiasm was backed up by ability, and Joe Lassiter quickly became one of the Company's stars.

In the face of strong bookings in the first half of 1974, d'Arbeloff became increasingly fearful that a replay of the 1970 scenario was in store. There were scattered signs of overbuying by semiconductor manufacturers, and the country's mood, reflected in a depressed stock market, was sour. Whatever d'Arbeloff felt about a given situation, he showed, and his sense of unease spread to his colleagues at Teradyne (Prestridge felt even more apprehensive than he did) and to any securities analysts he encountered. Teradyne stock, selling at $22 a share in June, quickly began a precipitous descent that saw it break $10 in October.

D'Arbeloff's premonitions were right, and bookings turned down sharply in July, as the semiconductor industry once again found itself with more manufacturing capacity than it needed, and as the U.S. struggled with a 12 percent prime rate. In the third quarter bookings totaled only $5 million after cancellations. To compound the gloom, the stock market kept falling, with the Dow Jones Industrial average below 600 for the first time in 12 years. Finally, the Nixon impeachment hearings were front-page news day after day until the President's resignation in August, sapping the nation's self-confidence.

As business collapsed, tensions mounted throughout the Company. Particularly hard hit were sales of linear IC test systems, as Detroit cut back on capital spending, and Linear Product Manager Graham Miller found himself on the defensive. Dick Lucey, Product Manager for Discrete Test Systems, could boast relatively good results but had to endure the common perception that his was a dying business. Board test business also turned down sharply, but here great hope was invested in newcomer Joe Lassiter, who soon became the third product manager for the line (the first was Tony Wakeling, who had moved to the U.S. from Europe; the second was Jack Jacobson, a native of South Africa and Teradyne's first Ph.D.). Yet Lassiter's quick rise also contributed to the tension, as other product managers began to resent the barrage of praise d'Arbeloff showered on his protégé. Lassiter, a major talent, hardly needed the boost, but in such matters d'Arbeloff was rarely subtle.

In Chicago, the pace quickened at Teradyne Central, where Russ Ashdown and Chuck Pearson were frantically searching for a way to market their line-test technology. In Boston, a Teradyne-funded venture named Camex was developing a computerized photo composition system that could prepare camera-ready advertisements for newspaper publishers. Once again, the technology was superb, the marketing strategy shaky.

And Milt Collins was back in the Teradyne fold. His Minico venture had failed, but Teradyne, whose entire inventory was now Minico-controlled, had little choice but to buy the Minico cadaver.

Trying to turn the lemon into lemonade, d'Arbeloff proposed a new Minico system to help electronics manufacturers manage wiring data — an application that could become an adjunct to the backplane test systems being marketed by brother George. Minico would eventually fade away, and Camex would later be sold, but Teradyne Central would eventually vindicate d'Arbeloff's faith in its prospects.

And what of Abe Bluestone, the erstwhile Teradyne sales manager who had joined Milt Collins in forming Minico? He fled the failing Minico to join Adar, a maker of memory testers. Adar was bought by Scientific-Atlantic, which then sold it to... Teradyne. The apple doesn't fall far from the tree.

A line of relay testers, managed by Ed Sutherland, showed brief promise, but as sales dropped to a trickle in 1974 the product group was merged into the Discrete Test Group, itself under a shadow.[1]

Despite the grim near-term outlook, the Company scheduled a meeting for securities analysts for November 12, at Boston's Pier 4 Restaurant. A larger-than-expected number of analysts (53) turned up to hear a series of talks by d'Arbeloff, Jim Prestridge, Joe Lassiter, Owen Robbins, Windy Hunter, Fred Macdonald, and Jack Salvador. The Company laid out a brutally candid picture of the situation, and the stock price fell from $11.25 to $8.50 within a week. Exactly two weeks later, the Company announced a furlough and salary reductions to deal with the downturn. But confidence in the Company's long-term future was unshakable. The last paragraph of the employee memo announcing the furloughs:

> "Hard times can be an occasion for gloom or an occasion for increased effort and determination to prevail. Gloom won't do us any good; a determined effort will. Teradyne has more going for it than any other company in our business, and the future is still ours if we earn it."

[1]The line was soon dropped, and Sutherland bought one of the remaining testers as he left Teradyne in 1975 to build a successful relay testing service in New Hampshire.

The furloughs alone were not enough to deal with the situation, and at the end of November 35 employees were laid off. Layoffs are never pleasant, but they were becoming commonplace as the economy sagged further. The press, of course, magnified every layoff announcement tenfold. The hiring of 350 workers was rarely worth a line of copy, but the termination of 35 was worth a headline.

The downturn could not have come at a worse time. The Company's rapid growth since 1971 had masked a number of problems which now bubbled to the surface. The Company's management structure was in fact structureless, with more than 20 division managers, product managers, functional managers, and staffers reporting directly to the President. There were two Vice Presidents, Jack Salvador having been given the title in recognition of his responsibility atop the Company's IC test organization, but no one could doubt that, whatever the titles, d'Arbeloff was de facto chief marketing manager and strategist for the entire Company. There was one exception: Teradyne Components enjoyed a fair amount of operational autonomy, suggested by the titles given to Windy Hunter and Len Johnson, the Division's President and Executive Vice President, respectively.

As bookings fell, resentments grew. The bicoastal reporting chain in semiconductor test (Miller reporting to Salvador reporting to d'Arbeloff) was difficult enough in good times; when bookings of linear testers were especially hard hit, the strains became acute. At Teradyne Applied Systems, General Manager Warren Cozzens had established Teradyne as the leading supplier of laser-trim systems, but this success produced its own strains. It had long been a point of pride at

Teradyne that the Company never promised more than it could deliver, and on innumerable occasions the Company had walked away from business on these grounds. Cozzens, however, was a consummate salesman, whose stock in trade was closing orders by making promises. It was easy, when business went sour, to blame either d'Arbeloff's conservatism or Cozzens's recklessness, depending on your point of view.

Despite the downturn and the tensions it produced, Teradyne in 1974 enjoyed a position any other ATE company could only envy. It owned more than half the market for digital IC test systems, more than half the market for linear IC test systems. It had more than half the laser-trim market, half the market for discrete semiconductor testers, plus respectable shares of the board-test and backplane markets. It was easy for an observer to conclude that, once the semiconductor industry started buying equipment again, Teradyne would continue to dominate its key markets.

For the year 1974, Teradyne posted record sales of $56.2 million. Like the record sales of 1970, however, the figure was deceptive, since shipments from backlog masked a mid-year collapse in bookings. The magnitude of the 1974 collapse would appear in the 1975 shipments, which, at $37.9 million, were down 33 percent from 1974. Despite losing a third of its revenue, the Company managed to show a nominal profit, thanks partly to Owen Robbins's deft asset management. In the President's letter in the Annual Report, d'Arbeloff offered this observation:

> "If we have learned how to stay profitable in the face of a 33% drop in sales, it's not a talent that we want to perfect."

Ignoring the woes of the semiconductor-test business, Teradyne Components recorded its sixth straight year of record sales in 1975.

The recession of 1974-5 would pass, and Teradyne's growth would resume in earnest, but grave damage was done.

A recurring debate in the ATE business was the optimum sizing of test systems. One can decide to market the smallest and most limited tester imaginable, at a low price, or the world's most ambitious test system, with a price to match — or anything in between these extremes. The art lies in finding the "sweet spot" on the price/performance spectrum, and it was an art that Teradyne of the 1960s had mastered. As the Company moved into the 1970s, it was occasionally seduced by the opportunities perceived at the low end, but at the high end, d'Arbeloff chose to let others chase the impossible. The market just wasn't there for the supertester. Besides, Teradyne could play that game only by compromising its hard-won reputation for reliability.

Thus the news that archrival Fairchild Test Systems was introducing a very aggressive new line of testers for VLSI (very large scale integration) ICs produced the same kind of derision with which Fairchild had responded to the J259 a decade earlier. "Given the number of parts in a Sentry," went the standard Teradyne observation, "and the MTBF (mean time between failure) of those parts, it's obvious that those Sentry systems will be down every hour." Downtime or no downtime, when Intel went shopping for a tester for its line of microprocessors, Teradyne was not in the market, and the Sentry won the coveted account. Teradyne had overlooked two points: First, to a customer in need of a tester, a part-time solution is better than no solution at all. Second, a customer like Intel, having made its purchase decision, will invest its own engineering resources to make sure that any system shortcomings are addressed.

The Sentry's resounding success had another unhappy effect: It revived a competitor that had been reeling and which, but for Sentry, might have exited the business.

The resurgence of Fairchild was not the only competitive threat on the horizon. In Japan, Takeda-Riken was hard at work developing a 100-megahertz test system, funded by the telecommunications giant NTT. Such blazing speed was more than anyone needed at the time, but Takeda's decision to seize the high end of the performance spectrum was worrisome.

Takeda-Riken, bankrolled by NTT as well as by part-owner Fujitsu, could well afford to fund the huge R&D effort required to produce its new tester. Teradyne, on the other hand, had only one source of R&D funding: its own sales and profits. There was no place at Teradyne for a tester that was not commercially viable — in other words, for a tester that could not be sold in quantity to a broad market. Budgetary limitations also meant that Teradyne had to pick and choose its projects, and at the top of d'Arbeloff's list was a tester for the fast-growing CMOS market.

Meanwhile, Teradyne's board-test business was showing signs of life, under Joe Lassiter. The main competitor was General Radio, the old-line instrument maker now building a position in ATE. The potential in a Teradyne-General Radio merger was too obvious to ignore, and an exploratory meeting was set up at a Waltham restaurant on April 13, 1972. Although a few more meetings were held and a formal merger proposal was developed some years later, it was not to be.

A number of other merger candidates were evaluated at this time, and some were the subject of fairly advanced negotiations. On several occasions renewed attempts were made to combine Fairchild Test Systems Technology and Teradyne, and intense meetings were held on both coasts, at the highest levels. For one reason or another, the discussions never bore fruit.

D'Arbeloff's view of acquisitions in high technology was decidedly negative, and this undoubtedly weighed against the successful completion of many negotiations. The odds against success were 10-to-1, he maintained, and history was certainly on his side. In time, Teradyne would make a number of acquisitions, most of which simply proved d'Arbeloff's thesis.

The exception, of course, was Teradyne Components, technically an acquisition but in reality a start-up. The Division was growing steadily on the strength of a product line that consisted of backplanes sold chiefly to the military. In those days customers would use specialized machines to wrap wire connections around the posts on these backplanes — a value that Teradyne undertook to add with the acquisition of a small wire-wrapping company called WrapCon (another acquisition that proved d'Arbeloff's rule).

As 1975 began, it was clear that the Company's morale was at rock bottom, and there was dark talk of a "major restructuring." Such restructuring was the subject of intense meetings during the week of January 6, attended by all product managers, Prestridge, Macdonald, Robbins, d'Arbeloff, and, from London, O'Connell. The mood was somber in the extreme, with d'Arbeloff and Prestridge especially pessimistic. Salvador assumed the mantle of semiconductor test leader (which, in fact, he was), and it was clear that Lassiter spoke for board test, with the other product managers (including linear test head Graham Miller) assigned to secondary roles. The gloom was palpable. The stock was selling at about $6, a major layoff seemed unavoidable, and a $1.2 million inventory write-down was to be charged against the fourth-quarter results. Happy days were gone again.

Those awaiting tangible results of the restructuring meetings did not have to wait long. On January 10, a bulletin board notice announced the creation of a new Manufacturing Systems Division (MSD), to be managed by Joe Lassiter and George d'Arbeloff. Lassiter, according to the memo, would direct product development, while the younger d'Arbeloff would lead the Division's marketing program and serve as Product Manager for the L100 Series of board testers. A young salesman named Jeff Hotchkiss would take George d'Arbeloff's place as Network Test Product Manager.

Alex d'Arbeloff always believed strongly in "participative decision-making," in which those to be affected by a management decision are brought into the process before the decision is announced. If some took this as management by consensus, they were mistaken. The d'Arbeloff approach gave the *illusion* of consensus, but there was no voting; the leader (d'Arbeloff) monitored the discussions, gained support for a decision that was unequivocally his to make, and effectively preempted any opposition.

Because of this strongly ingrained management principle, an organizational move, once announced, was carved in stone, not subject to recision. But the January 10 memo, tellingly, was not issued over Alex d'Arbeloff's signature, but instead was signed by the two new Division managers. Whatever the level of "participative decision-making" behind the memo, it was rendered inoperative only a month later, when it was announced that George d'Arbeloff was leaving board test to consider a new assignment in field service.

Lassiter chose as his Division's new Engineering Manager John Turner, the English engineer from Mark Instruments, and Turner mounted an effort to develop a new board test system capable of testing boards having analog as well as digital circuits. Meanwhile, the new Division would continue to offer the

L115 digital board tester and the L427 Shorts Tester, a $10,000 product designed to detect flagrant manufacturing defects (e.g., short circuits) as part of an attempt to construct a product line that spanned the entire board test process. Then, of course, there were the network, or backplane, testers, which had found a niche market among makers of telecommunications equipment. With the able and affable Jeff Hotchkiss directing marketing, and with Lassiter leading the Division, there were solid grounds for hope.

The semiconductor test operations were another matter. This was where the industry downturn was felt most severely, and the Company scrambled to generate business wherever it thought it sensed opportunity. A small Massachusetts company called Accutest produced benchtop testers for calculator chips, watch chips, and low-end linear ICs, and Teradyne agreed to sell these under the Teradyne name. A Company-designed microprocessor test system, the J199, was offered either as a standalone system or as an add-to the Teradyne J384 Memory Tester.

The Company had brought the J384 to market two years earlier, targeting Intel, the pioneer in the memory business and the hands-down market leader. In 1975 the system was upgraded to handle the new 4K random-access memories due from Intel. Later a Dallas chip maker, Mostek, would come to dominate the 4K DRAM market and would be the first to introduce a practical 16K part, and Mostek's success generated demand for more Teradyne memory testers. The years ahead would certainly produce a demand for thousands of memory testers, the kind of quantities that spelled huge profits. Clearly, the memory test business represented an enormous growth opportunity for Teradyne, and the Company deployed its engineering resources accordingly.

The Company's flagship IC test system, the J283 "SLOT" Machine, had scored a direct hit on the market, with more than 200 systems shipped since its introduction, but it was time to move on, and the J283's successor, the J325, was being readied at Chatsworth. The J325, essentially a solid-state, and therefore faster, version of the J283, would be sold in two basic configurations — one equipped with a high-voltage test station for CMOS circuits, the other with a low-voltage station for bipolar circuits. (The distinction between bipolar and MOS testing was starting to break down.)

The linear IC test line was anchored by the J273, the successor to the successful J263. The J273 was also a winner, with more than 200 shipped, but it was not a winner in the 1975 recession. A derivative system, the J279, was introduced at the IEEE Spring Conference in Chicago in June. Its intended market was makers of radio and television sets, who could use the system to test and align their circuit boards quickly and automatically. Linear-test product manager Graham Miller was fascinated by the possibilities in consumer and automotive electronics, but d'Arbeloff was less enthusiastic.

The recession also took its toll on the Company's laser trimming systems, and the Company responded by introducing a new lower-priced system, the W211, in an effort to bolster sagging sales.

Only one of the Company's businesses escaped the onslaught: Teradyne Components, whose chief customers were military contractors. The cold war was still raging in 1975, and Teradyne Components' backplanes were finding their way into a broad range of weapons systems. It must be added that the Hunter-Johnson-Neavitt troika gave the Division a management capability second to none in that business.

On May 15, 1975, d'Arbeloff, accompanied by Prestridge and Robbins, journeyed to New York for the Company's second appearance before the New York Society of Securities Analysts. As in 1971, the timing could hardly have been worse. Business was terrible, and there was no attempt to sugar-coat the fact. In such forums, in fact, d'Arbeloff tended to overstate the magnitude of the problem. If problems lurked, he wanted everyone to know about them. But this time there would be no talk of customers "shipping junk." That lesson had been well learned.

The only concession made to the analysts' appetite for a "story" was a broad-brush report on the progress of the 4TEL product. D'Arbeloff shied away from market projections, but in the question and answer session the analysts persisted. If there were 80 million telephone lines in the United States, and if Teradyne could glean even two dollars a line for 4TEL, might that not prove a bonanza? But d'Arbeloff would not bite, and the analysts were left to dream on their own.

D'Arbeloff, in fact, was a master at dissembling. When an analyst bore down on a delicate subject, he would become the artful dodger, insisting on answering a different (and easier) question. Thus:

> Analyst: Have you sold any of these systems to the Bell system?
>
> d'Arbeloff: That's not the question you should be asking. What you should be asking is whether we have a long-term strategy to become a leading supplier to the telephone industry. Let me tell you what we're doing to make that happen...

As 1975 unfolded, orders remained disappointingly weak. First-half bookings were fully 60 percent below the prior year's first half. The new digital/analog board test system (called the L125) was not moving, nor was the J199 microprocessor tester. And, despite an energetic sales effort, at mid-year not a single 4TEL system had been sold since the trial unit was installed at Sun Prairie.

As the Company floundered, further management shifts were made. Long-time Operations Manager Fred Macdonald was moved into the Human Resource area, while Controller Bill Siddall took over Operations. Since Siddall was hired by and reported to Robbins, this appointment essentially reinforced Robbins's role as the Company's operations chief.

Much as d'Arbeloff craved diversification, it was clear that Teradyne really understood only two markets — ATE for electronics and backplanes. (In time, telecommunications would be added to the list) — and that the world beyond was *terra incognita*. Camex, in d'Arbeloff's mind, proved the point, as he grew increasingly impatient with what he saw as its marketing deficiencies. The venture's photo composition system embodied excellent technology; of that there was no doubt. The Camex system's debut at a Houston trade show in June was a spectacular success, and The Boston Globe declared its intent to buy. But Teradyne was skirting perilously close to break-even, and d'Arbeloff's support of the 4TEL venture was using up much of his capital with his Board. Something had to give, and Camex, obviously outside Teradyne's core competence, was the leading candidate. The search began for a buyer.

An upturn in bookings in August kindled hope that the recovery was near, and previously imposed salary cuts were restored for all but the highest-salaried employees. More significantly, United Telecommunication, a major U.S. independent telephone network, agreed to purchase 4TEL systems for its service centers in Florida and North Carolina. The gloom that had enveloped Essex Street began to lift. But within a month, it was crisis time again.

Teradyne test systems were used on the production lines of many of the world's leading semiconductor manufacturers, and it was axiomatic that these manufacturers were hostage to the reliability of their Teradyne systems. The silicon chips are worthless until the test system judges them to have value, and an inoperative test system is nothing short of a calamity. Indeed, this was the premise on which DeWolf and d'Arbeloff built a company and an industry.

Thus, when reports from the field indicated a surge in test-system failures, the alarms sounded at Essex Street. After analyzing the failure data, Teradyne engineers gave d'Arbeloff the bad news: An integrated circuit used in Teradyne's memory testers, board testers, and computers was failing at a disturbing — and accelerating — rate. The IC, a low-power gate, was supplied by National Semiconductor, which, to complicate matters, was a major Teradyne customer. Eyeing the possible damage, d'Arbeloff began considering the possibility, rapidly becoming the probability, of having to replace every suspect circuit board in every system shipped since mid-1974. The cost: at least $2 million, enough to plunge the Company into the red. But there was no alternative, as the Company's reputation was squarely on the line. The atmosphere at Essex Street, never what one would call relaxed, was supercharged as a strategy was devised for handling the disaster.

Fortunately, relations between Teradyne and National were good, and after some spirited negotiations the two companies agreed on a plan to share the costs of the retrofit. D'Arbeloff's negotiating skills, which were and are legendary, were never put to better use.

14

• • •

The Blow-Up of 1976

As 1976 began, orders strengthened markedly, and the stock rocketed from $10 to $20 in the first five months of the year. The creation of two Teradyne ATE "super-groups" was noted by *Electronic News* in its February 9 issue. Presiding over the semiconductor test universe was Jack Salvador, while Lassiter headed the board test division. Reporting to Salvador were Graham Miller as Division Marketing Manager, Bill Attridge as Technical Director, Wayne Ponik as Logic Test Product Manager, Phil Phillips as Memory Test Product Manager. and Dick Lucey, Product Manager for Discrete and Linear IC test systems. Thus the bicoastal reporting chain became even more convoluted, with west-coast product managers Ponik and Phillips reporting to Miller on the east coast, who reported to Salvador on the west coast, who of course reported to d'Arbeloff in Boston.

Reporting to Lassiter in the newly formed Manufacturing Systems Division were Jeff Hotchkiss (Backplane Test Product Manager), Fred Macdonald (leaving Human Resources to become Board Test Product Manager), and, in Chicago, Warren Cozzens (Laser Trim Product Manager).

The reorganization had the desired effect of greatly reducing the number of people reporting directly to d'Arbeloff, but there was inevitably some second-guessing, as the recent recession had left scars. Cozzens reacted to the reorganization by leaving the Company. But with business improving weekly, most resentments were put aside — or so it seemed.

A conversation with National Semiconductor's Pierre Lamond in 1975 convinced d'Arbeloff that conventional IC test systems suffered from two major limitations: high costs and the difficulty of writing test programs. Both limitations, Lamond argued, could be overcome through the use of microprocessor-based systems. D'Arbeloff, sold on the concept, tried to interest the existing semiconductor test organization to mount a project, but that organization, mired in its own projects, was cool to the idea. D'Arbeloff then hit upon a novel plan: A new, ad hoc group would be set up specifically to develop the new tester. Technically, the group would be part of the Semiconductor Test Division, but only technically. The group would be in Boston, and its sponsor would be Alex d'Arbeloff. Bill Attridge was asked to lead the development, and Tom Newman, who had joined the Company in 1973, was pulled in from the New Jersey sales office to become the product manager.

Years later, IBM would mimic d'Arbeloff's "off-campus" approach when it sited its personal-computer team in Florida, far removed from its ongoing computer operations. And years after that, d'Arbeloff would repeat the scenario in placing Teradyne's J750 "Integra" product outside existing reporting channels, with stunning success. The 1975 version of the story would end in failure, however.

The product, called the J401, was conceived as a low-cost, benchtop tester in which programmed parameters and limits could be entered on a keyboard and small screen. A number of compromises were made in order to keep the cost down. The real art in test system design always involves the choice of compromises, but in this case a poorly defined target market added to the challenge. Was the customer to be the semiconductor manufacturer, or was the system aimed at the incoming inspection market? If the former, it was vital that test results correlate with those from the large test systems on the production floor, but here the J401 was at a disadvantage, for it lacked the Kelvin connections necessary to eliminate the effects of handler or prober cables. If the latter, it was essential that the tester cover the widest possible range of devices, but ease of programming had been purchased at the expense of flexibility. Most damning of all was the system's inability to test CMOS devices. Also, the system's price of $40,000, while far below that of large production test systems, was higher than most incoming inspection budgets could swallow.

Newman and the sales force gave the J401 their best shot, but after a year only a dozen systems had been sold. The next year, Teradyne's Japanese operation, by adding a second test station to the system, was able to sell another dozen. Then the system was given a quiet burial. Once again, Teradyne had proved that selling "80 percent of the performance for 50 percent of the price" was a bad bet. The customer always knew too well which 20 percent of the performance you left out.

At the beginning of May, 1976, two men were the clear leaders of the Company's semiconductor test efforts: Jack Salvador ran digital testing, and Graham Miller was the linear czar. Two months later both would be gone. The shock would shake Teradyne to the core and would reverberate for years thereafter.

The proximate cause of Salvador's dismissal was a disagreement over engineering priorities at Chatsworth. At issue was a promise made by d'Arbeloff to Intel's Andy Grove some months before. Grove was incensed over the unreliability of Teradyne memory testers on Intel's production line, and he let d'Arbeloff know it in no uncertain terms.

"But Andy," explained d'Arbeloff, "your people keep demanding changes in the system, and if we can't stabilize the design we can't straighten out the reliability issues." Grove knew the argument was valid, and he agreed to review the current list of Intel-directed changes, 19 in all, and then eliminated all 19. In return, he extracted a vow from d'Arbeloff that the reliability problem would be solved — quickly. D'Arbeloff immediately flew to Chatsworth and told Salvador that all else must be subordinated to the task of addressing the system reliability issue.

Returning to Chatsworth a few weeks later and finding Salvador's engineers busy designing a new system option, d'Arbeloff fired one of his two Vice Presidents.

The full story, of course, had to have been more complicated than that. Even assuming the worst, the Intel flap on its own could not have been worth the explosion and its aftershocks. A cultural divide between d'Arbeloff and Salvador had been simmering for some time. D'Arbeloff's Teradyne was a perkless society in which there were no office doors and everyone flew coach. Salvador flew first class (the better to work on the plane), and his office was guarded by a protective secretary. Salvador's manner was measured and cool; d'Arbeloff was volatile and could be harshly direct. It would be hard to find two people whose temperaments were so completely mismatched. (On the other hand, one might have said the same of d'Arbeloff and Prestridge, who worked so well together. Human relations are not always easy to parse.)

To the credit of both parties, no effort was made to whitewash the firing, either in the press or in the May 24 announcement to employees, which began with the following:

> "As a result of a disagreement with me over the longer-range strategies to be followed by Teradyne, Jack Salvador has left the Company. Such honest differences are by no means uncommon among those responsible for managing a company's affairs, and where they are irreconcilable, a parting of the ways is sometimes the only answer, no matter how regrettable it may be on a personal level."

The memo went on to say that d'Arbeloff himself would take over as acting manager of the Semiconductor Test Division, but almost immediately he announced the appointment of a new hire, Ed Rogas, as Assistant Division Manager. Rogas, formerly with American Research & Development and before that a staffer at Cabot Corporation, would, initially at least, be based in Boston, at d'Arbeloff's side. Rogas undoubtedly came on the strength of recommendations from d'Arbeloff's ARD contacts, and his background, like Prestridge's, included degrees from the Naval Academy and the Harvard Business School. His lack of industry experience was an obvious handicap, but d'Arbeloff believed he would be a quick learner.

Speculation naturally arose as to the reaction of Graham Miller. It was clear that linear IC testing was not only a fast-growing business in its own right, but was a gateway to sales in the automotive, telecommunications, and consumer electronics industries. Despite repeated attempts, Fairchild had failed to mount a successful competitive attack, and Teradyne in 1976 virtually owned the market for linear IC test equipment.

The speculation ended in mid-June with the news that Miller was leaving Teradyne. Worse, he declared his intention to take six of his best engineers with him to found a competing company. In the wake of the Salvador departure, the news was devastating.

Graham Miller was an extremely able manager, and he was well liked at Teradyne. He had come to the United States with Teradyne's acquisition of Mark Instruments, and he had been largely responsible for making Teradyne the unquestioned leader of the linear test business. In the process, he developed excellent relations with key customers throughout the industry. He would obviously use these relations to advantage in getting his new company off the ground. Even more ominously, by gutting the linear engineering group at Teradyne, he was crippling the Company that would be his major competitor.

Had Miller reacted impulsively when d'Arbeloff snubbed him by appointing outsider Rogas to run semiconductor test? Almost certainly not; one does not plan a new company overnight. Graham Miller and his engineers had to have been quietly laying their plans for some time. The opportunity was no doubt irresistible: Teradyne thoroughly dominated a market that seemed certain to explode. Someone, someday was certain to mount a successive competitive threat. Why not Miller?

Another factor in the mix was Miller's strong interest in the automotive and consumer-electronics markets, an interest not shared by d'Arbeloff. Thus, when Motorola invited Teradyne to develop systems for testing AM-FM radio receivers, Teradyne declined, with d'Arbeloff overruling Miller's desire to commit engineering resources to the project. Miller also made sure that the key engineers working on consumer electronics were present when d'Arbeloff delivered his put-down.

D'Arbeloff seethed at Miller's defection, as he had thought he had Miller's promise to remain aboard until the Salvador aftershocks subsided, but there was little to be done except to try to rebuild the linear group. Linear IC testing had represented 30 percent of the Company's first-half test bookings, and the loss of that business would be hard to swallow.

Fortunately, linear IC test orders remained relatively strong throughout the year, thanks to the momentum established by the J273. Eventually Graham Miller, intimately familiar with the J273's weaknesses as well as its strengths, could be expected to unveil a strong competitive threat. The challenge at Teradyne was to prepare for that threat by developing a next-generation system.

To direct that effort, John Turner left his post as board test engineering director to become manager of the linear test group, which was renamed the Industrial/Consumer Division. Turner almost immediately began laying plans for a new breed of linear tester with upgraded digital capability. The new system would be known as the A300 series, and Turner coined a new term to describe Teradyne's assault on the mixed-signal test market: The A300 would be an Analog LSI Test System.

Institutional investors were somewhat rattled by the departures of Miller and Salvador, but business at Teradyne was strong, and the stock price responded to that fact rather than to the personnel turnover. Sales for the year totaled $53.7 million, up an astonishing 42 percent over 1975. In an attempt to show the world that Teradyne could still field an outstanding management team, the 1976 Annual Report featured a series of interviews with various executives, including Jim Prestridge, Owen Robbins, Bill Siddall, Ed Rogas (now managing the Semiconductor Test Division), Bill Attridge, Joe Lassiter, John Turner, Chuck Pearson, Russ Ashdown, Windy Hunter, and Len Johnson. All in all, it was an impressive array of talent — arguably the strongest lineup of any $50 million company anywhere.

Early in 1977, less than a year after installing Rogas atop the Semiconductor Test Division, d'Arbeloff realized that his newest manager was struggling, not because of any lack of management ability or "smarts" — Rogas had plenty of both — but because the magnitude of the job demanded someone with experience in the test arena. The pot was also stirred by John Turner's attempt to staff his new analog LSI project with digital engineers from the California operations.

D'Arbeloff moved decisively. He asked Jim Prestridge to move to California to replace Rogas as head of the Digital Test Division. Dennis O'Connell, European Sales Manager for the past six years, was then moved from London to Boston to succeed Prestridge as world-wide Sales Manager. O'Connell, whose leadership ability was matched by an unsurpassed knowledge of the world semiconductor scene, was also an English gentleman in the best sense of the term, and his arrival brought to Teradyne's headquarters a civility and courtliness that, it must be admitted, were not always the Company's strong suits.

One of Prestridge's first moves as Division Manager was to persuade Rogas, who might have been expected to bolt, not only to stay with the Company, but to move with him to the west coast. It was a brilliant stroke by a master strategist. Prestridge and Rogas would team up to repair any wounds left by the Salvador departure and to move Teradyne quickly into the VLSI test arena. Rogas would then go on to manage the memory test line and eventually to become one of Teradyne's senior officers. Rogas was quick-witted and intensively competitive, peppering his conversation with well-placed barbs followed by staccato bursts of laughter. Like many of his Teradyne colleagues, Rogas did not suffer fools gladly.

Turner, rebuffed in his efforts to annex talent from the digital division, hired the engineers he needed and got on with the task of creating the A300. The storms of 1976 had been weathered, and the good ship Teradyne was sailing on.

Industry observers who noted Teradyne's recovery from the major shocks of 1976 might have concluded that here was another example of d'Arbeloff's toughness under adversity, and that is partly true. But d'Arbeloff himself knew where most of the credit belonged. In Salvador and Miller, Teradyne lost its core business's two principal line managers, with responsibility for all of the Company's digital and linear IC test lines. Their departures left a leadership vacuum that only one person in the Company could fill. There were other talented people on d'Arbeloff's team, but in his knowledge of the business and his management ability Prestridge was in a class by himself. He was what one calls a "quality person," and d'Arbeloff knew only too well what a treasure he had stolen from Hewlett-Packard seven years earlier.

"In those days," d'Arbeloff reflects, "when I dealt with the various division managers, I always felt that I was teaching them.

"But the truth of the matter is, Jim Prestridge was teaching me."

15

• • •

Back on the Growth Track

In 1977 it was clear that Teradyne was once again on the move, as sales soared to $67.8 million, a 26 percent gain over 1976. The profit picture was even brighter, as net earnings more than doubled, to almost $5 million. The 4TEL system was now installed by units of two major independent telephone systems (GTE and United Telecom), both of which had approved the system as standard equipment. The J325 was well on its way to becoming one of the Company's major successes, with more than 200 systems shipped, and it was joined in the logic test market by a companion system for pulse parametric testing, the S357. But the lack of a test system for microprocessors was painfully obvious.

The J387 Memory Test System was a huge success, in those days when the United States was still the hub of the world's memory business. The growth of the logic and memory test lines exhausted capacity at Chatsworth, and the Division looked for new space in the San Fernando Valley. The linear test group rebounded after the loss of Miller and his engineers and upgraded the J273 into a high-performance, -B version. There was still a respectable revenue stream from the transistor test line, and film-resistor and hybrid-circuit manufacturers were buying a new laser-trimming system, the W411.

The Manufacturing Systems Division introduced a low-cost board tester, the L529, which featured a clever fixturing system, in which the board under test was pushed against the test contacts by a matrix of precision-weighted rods that applied uniform pressure over the board's surface. The Division's flagship product was the L135 Board Test System, whose blinding speed (5 MHz!) and 444-channel capacity prompted Lassiter to call it an "LSI" Board Test System. Two network testers — the man-sized N151 and the bench-top N123 — offered manufacturers an easy way to test wiring networks.

Teradyne Components still derived much of its sales from military contractors, but the commercial market, trying to keep pace with the formidable packaging requirements demanded by LSI circuits, was beginning to consider Teradyne's "two-piece" approach, in which a pair of multicontact connectors — one on the backplane or motherboard, one on the daughterboard — offered a level of performance and contact density that was impossible to achieve with traditional one-piece systems.

The loss of Salvador in 1976 had left the Company with only a single Vice President — Prestridge — but 1977 brought a bumper crop, including Joe Lassiter (MSD), Dennis O'Connell (Sales), Owen Robbins (CFO), Bill Siddall (Operations), and John Turner (ICD). D'Arbeloff's "first team" was thus firmly in place, and they were ready to take the Company to heights unimaginable back in the gloom of 1975 and 1976.

The year also saw some turnover on the Board of Directors. Nick DeWolf, now living in Aspen, decided to call it a day, and Greylock's Bill Elfers also decided to leave the Board. Their replacements were Dan Gregory of Greylock and Walter L. ("Jake") Lingle, a retired Procter & Gamble executive.

Good as 1977 was, it was only a taste of what was to come. Throughout 1978, bookings and shipments climbed steadily. Earnings per share bounded from 39 cents to 52 cents to 71 cents. Through October, the stock followed suit, doubling from $15 to $30. In the last two months, a wave of selling brought the price down to $16, some institutions no doubt feeling that, given the volatility of the business, the good times were due to end soon. In this case, they were dead wrong.

In California, Prestridge found his new plant site in Woodland Hills, just a few miles south of the Chatsworth plant, now groaning under the weight of record shipments in both memory test and logic test. The memory test line, now with Ed Rogas as product Manager, had that market's wind at its back, and it poured on the engineering, adding capabilities like Real-Time Bit Mapping, or RTBM. While stressing the memory, the user could watch bits fail, one by one, on the color RTBM monitor.

In the new Industrial/Consumer Division, Turner recruited new engineers at a rapid pace. Graham Miller's LTX had emerged as a viable competitor, but Teradyne was able to use its larger revenue stream to fund more ambitious efforts. College recruiting was always a top priority at Teradyne, as the Company's Human Resources staff, accompanied by engineers who had graduated from the target colleges, would fan out each spring to 30 or 40 colleges around the U.S. — a program that was maintained whatever the state of business.

In board test, the new L135 Test System found quick acceptance, particularly among makers of telecommunications equipment, and the network testers and the L529 were also doing well, prompting expansion of the Division's facilities in the Lincoln Street building.

Users of board test systems chronically complained about the massive effort required to write test programs, and these complaints spelled opportunity to Lassiter, who engineered the purchase of a small Texas company called Digitest in order to gain control of a software package called D-LASAR, which could be used to help generate programs automatically.

Thanks to GTE and United Telecom, more than 700,000 telephone lines were being regularly tested by Teradyne's 4TEL system as 1978 drew to a close. But the division suffered a jolt with the departure of Chuck Pearson, its sales and marketing leader, who had joined Russell Ashdown to launch the whole 4TEL effort in 1973. Since then, the two had served more or less as co-managers of Teradyne Central, with Ashdown focusing on engineering and Pearson directing sales and marketing. Both were extraordinarily gifted managers, whose talents were clearly complementary, and they functioned as partners very effectively for several years. But, as d'Arbeloff (of all people) must have known, two people cannot run one business; sooner or later one of them will leave. In this case it was Pearson who left, and Ashdown was named a Vice President a year or so later.

All in all, the Company in 1978 was in almost frighteningly good shape. Backlog was high, the balance sheet solid, the products all selling well, the management team working well together, and d'Arbeloff back in stride. The upheavals of 1976 were now a distant memory. What could possibly go wrong?

The answer, in 1979, was: "Nothing." Sales shot through the $100 million threshold, with bookings and year-end backlog also setting new records.

A tidal wave of new orders hit the Company right off the bat, in January, and the torrent continued throughout the year. Some of the good news could be attributed to a general upturn

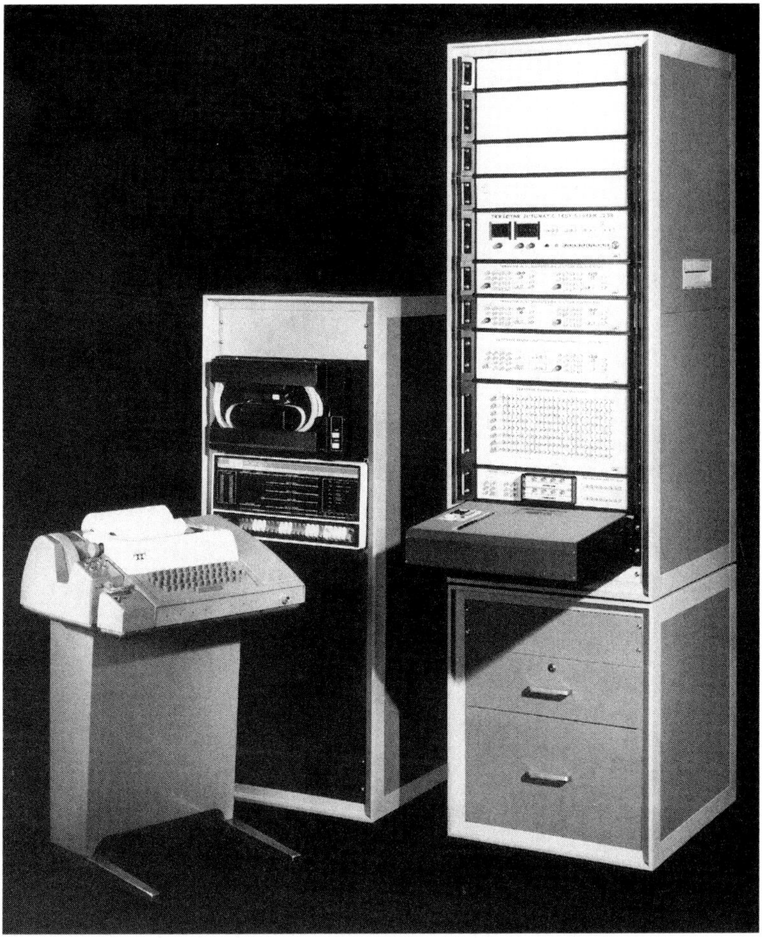

The J259, industry's first computer-operated IC test system.

Nick DeWolf (right) and Digital Equipment Corporation's Ken Olsen at the J259. On the strength of the J259, Teradyne quickly became one of DEC's largest customers.

The Teradyne Follies' Chorus Line, The Terabelles.

Arlene Harrington as "The Elevator Girl" in the Follies. Joe Laquidara is on guitar, Ed Sutherland plays the banjo.

Open House for families at the Essex Street plant. No attempt was made to hide the pipes and pillars and ducts. They were instead painted in garish colors.

Kemon Taschioglou.

Joe Lassiter.

Russell Ashdown.

Jim Prestridge.

Owen Robbins.

A multi-station transistor tester of the 1970s. The distinctive "kiosk" rack was designed by Walter Kern. The system's computer, in the second kiosk from the right, is a Teradyne M365.

The Teradyne look of the 1980s featured this streamlined package designed by Paul d'Entremont. The system shown is an L290 Circuit Board Test System.

Alex d'Arbeloff and Dennis O'Connell.

Windy Hunter.

Len Johnson.

Teradyne's three sales managers discuss the world of ATE. Left to right, David Hodgson, Europe; Rick Dyck, Japan; John Casey, North America.

The Harrison Avenue Building — before and after.

in semiconductor equipment buying, including heavy test system buying from the Japanese. But the Company also made its own good fortune by managing itself with great skill. The team d'Arbeloff assembled coming out of the 1975 recession and the 1976 management upheavals was now operating smoothly and effectively, making all the right strategic moves.

For example: For several years the Industrial/Consumer Division had been developing its new Analog LSI test system, called the A300. The A300 was the first post-Miller test system, and it looked strong on paper. There was, however, one serious problem: As customers heard whisperings about the coming A300, they might put off purchases of the current system, the J273B. The J273B was a major contributor to Teradyne's sales, and the Company could ill afford to lose these sales in the months and quarters before the A300 was ready. Solution: Arm the J273B with an array of new capabilities that would make it irresistible. The J273B was also promoted heavily by a "3-C" (Capability, Commitment, Compatibility) campaign, with a brochure that paraded photos of 54 of the Division's engineers (lest anyone think that the exodus of Miller and company had left the group talent-starved). The campaign was successful, and J273B sales continued to build even after the A300 was announced at a trade show in May.

The A300 was to become one of Teradyne's all-time best sellers, with more than 1800 systems shipped. The new Analog LSI Test System straddled the analog and digital worlds, featuring a much stronger digital capability than its predecessor and an architecture that accommodated the wide variety of instrumentation needed to test analog and mixed-signal devices.

Logic system sales remained strong, with the old SLOT Machine refusing to fade away, even with competition from the newer, faster J325. In memory test, shipments of the J387 passed the 200 mark in mid-year. This was the heyday of memory testing at Teradyne. Some 435 J387 systems would eventually be shipped. This total, added to 365 J386 systems and 215 J385s, gave Teradyne a wide lead over all other suppliers. More important, customers tended to buy memory testers by the dozen rather than singly, so margins were excellent. The cost of selling and supporting 50 memory testers to a single customer is obviously far less than the cost of selling and supporting 50 testers to 50 different customers. High volume also allows one to amortize engineering costs quickly, before the system becomes technically obsolete. In the ATE business, no money is made on the first 20 or 30 systems, but a great deal of money can be made on the 100th or 200th system. Thus the memory test business, with its promise of high unit volume, remained near the top of Teradyne's priority list despite demand volatility. Engineering programs were well funded, and a long list of new features, including Automatic Edge Control and more advanced Real-Time Bit Mapping, were added to fortify Teradyne's position in the market.

Unfortunately, other well funded engineers in Japan were also hard at work.

With the number of Teradyne test systems at a single customer site often running into the dozens, and with each system generating a sea of test data, there was an obvious need for systems to manage the systems — to collect and analyze the test results, to record system parameters such as downtime and throughput, and even to serve as sites from which test programs could be downloaded on demand. Such master computers were in fact often designed by the customers themselves. Seeing a ready market, Teradyne produced several such systems, sold under the names of Test System Administrator (TSA), Test System Director (TSD), and Test System Manager (TSM). In the most common configuration, the product was a Digital Equipment PDP-11 computer plus the necessary command and control software. Hundreds of such systems were sold (in fact, many A300-series testers were useless without them), and a new product designation and organization, the Host Products Group, was formed under Chip Thayer, a former Teradyne salesman.

Expansions were underway at all Teradyne sites. In Nashua, Teradyne Components, staring at a 50-percent increase in sales for the year, broke ground for a new 140,000-square-foot headquarters. In Woodland Hills, a 50,000-square-foot addition was under construction. In Boston, where the Essex Street and Lincoln Street buildings were both fully utilized, Teradyne exercised its option to purchase the latter for $2.4 million, then took out a $9.3 million mortgage on the building. The Company, whose real-estate subsidiary was called, tongue in cheek, "Terafirma," was proving especially adept at ferreting out distressed downtown real estate at give-away prices, then turning them into workable industrial space.

As the Company grew at Woodland Hills, Nashua, and elsewhere, the share of total employment belonging to Boston dropped steadily. The Boston sites brought with them two problems that were never adequately solved: The first was lack of suitable parking facilities. The second was the reluctance of many professionals to commute from their suburban homes each day into the inner city. Some young engineers fresh out of school found the city life style inviting, but most recruiters saw the Company's headquarters location as a minus. From time to time someone floated a suggestion that the Company move to the suburbs, but d'Arbeloff would hear none of it. In this he was stoutly supported by Owen Robbins, to whom the city offered pleasures never tasted in Pauls Valley, Oklahoma.

Two Vice Presidents were appointed in 1979: Russ Ashdown, whose 4TEL system was now on line testing more than 1 million telephone lines, and Avak Avakian, brought in from GTE Sylvania to help manage Operations. The Company, which had somehow survived most of its first 16 years with one Vice President, and never more than two, suddenly had eight.

At the 1979 edition of the Wescon trade show, Teradyne unveiled a new test system that offered engineers an innovative, powerful way to test and trouble-shoot circuit boards. The L621 was in effect a self-contained engineering laboratory, a computer-controlled mainframe and communications console that could orchestrate a wide range of plug-in signal sources and measuring instruments, all available from Teradyne. The concept, which years later would be popularized by others under the heading of "virtual instrumentation," proved highly popular with Teradyne's own engineers, and soon the L-shaped L621 consoles were seen throughout the Lincoln Street plant, where engineers manipulated arrays of sine-wave generators, voltmeters, oscilloscopes, counters, and other instruments without budging from their chairs. The L621 was a real "first," which deserved a better fate than it got, but the Company simply did not understand how to sell into the laboratory market, where Hewlett-Packard reigned supreme. The L621 died a slow death, but its memory is still cherished by many.

Teradyne was now a $100 million company, a leader in the world markets for its products, and a force to be reckoned with. There were still competitors — spin-off LTX was building a solid position in mixed-signal testing, and Fairchild was back in stride on the strength of its Sentry VLSI test systems, which had still not evoked a competitive response from Teradyne. There were other, smaller competitors, and the Japanese had to be watched, but in general, it was fairly smooth sailing for Teradyne as the 1970s gave way to the 1980s.

On May 15, 1979, Teradyne stock was listed on the New York Stock Exchange (since 1970 it had traded on the American Stock Exchange). The first 100 shares to cross the NYSE tape were bought by Alex d'Arbeloff, for $19 a share. Seven months later, the Company raised $18.6 million by selling 550,000 new shares at $35.75. The entire offering was sold out the day it came to market.

The two hot products in 1980 were the J387 Memory Test System and the A300 Analog LSI Test System, both shipping at a one-a-day clip for most of the year. The memory test situation was especially buoyant, with Mostek's plant in Texas alone accounting for about 100 systems. The system was also on the test floor at an NEC plant in Ireland, where its performance drew praise from the factory personnel. Unfortunately, it also caused NEC's Tokyo executives to decide that the memory test business was too important to leave to the Americans, and before long an NEC subsidiary, Ando, had its own testers on the same Irish test floor. The Teradyne systems were preferred for their reliability and throughput, and they were also favored by NEC's Irish technicians, to whom Boston was a more congenial training site than Tokyo, but Ando's NEC connection would ultimately carry the day, at that plant and elsewhere.

Progress in the board test area was agonizingly slow. General Radio, now doing business as GenRad, had sold off most of its old instrument lines and was wholly focused on its board test business. Fairchild fielded an acquired board test line as well as the Sentry IC testers. At Teradyne, the board test business was a distant second to semiconductor test in sales, with 4TEL and Teradyne Components, now renamed Teradyne Connection Systems, coming up fast. It was clear that the Company's L135 Functional Board Tester was incapable of taking meaningful market share away from GenRad, and Lassiter, with d'Arbeloff's blessing, decided to throw a "Hail Mary" pass.

Teradyne's chief entries in the board test arena were the L125 and, later, the L135 functional test systems. The Company also marketed, with less success, an in-circuit tester, the L529. The two approaches — functional and in-circuit testing — represent two different ways of addressing the board test problem. In functional testing, the tester checks the outputs of the board while the inputs are exercised. In in-circuit test, the system contacts the board's innards through a mechanical fixturing system in an effort to find manufacturing defects (missing components, solder shorts, etc.). Functional testing requires a large investment in software, is more expensive, and delivers higher fault coverage. In-circuit testing is simpler (except for the fixturing), less expensive, and is best used in a screening operation designed to weed out egregious faults before the board is sent to functional test. Lassiter's new system, called the L200, began life as a more powerful edition of the L100 series, but the objective soon became much more ambitious. The L200 would combine *both* primary board test technologies — functional testing and in-circuit testing — in a single mainframe.

The L200, whose architecture was the creation of a young engineer named Dave Sulman, at first seemed impossible to build and test. Its main circuit boards, or "channel cards," were larger than anything ever attempted at Teradyne (or almost anywhere else, for that matter): 15 by 20 inches, a size dictated by the fact that each had to accommodate 300 integrated circuits. (The rule of thumb was one square inch per IC.) A visitor from Digital Equipment took one look at the Teradyne channel card and declared it utterly impossible to produce.

Not only would the new board tester be a manufacturing nightmare; it would also sport a high price tag: about $1 million, or four to five times what the market was then paying for functional testers. Thus Lassiter was essentially betting the Division, not to mention his career, on his ability to sell the new system. D'Arbeloff, who years before had doubted the marketability of another megatester — Fairchild's Sentry — was a solid supporter of the L200. It was hard to be otherwise in the face of Lassiter's irrepressible enthusiasm. The system was introduced at a trade show in the fall, to gasps from customers, editors, analysts, and competitors. The consensus: No one, but no one, would ever pay a million dollars for a board tester.

For the better part of a year, it seemed that the doubters were right. Then MSD's Jeff Hotchkiss closed a sale at Kodak. Within another year, the L200 was the talk of the board test business.

Sales of $165 million in 1980 meant that the Company had achieved a five-year compound growth rate of 34 percent, and Teradyne was added to Fortune's list of the 1000 largest industrial companies in the United States. There would be bumps along the way, but after 20 years Teradyne's place in the sun was secure. The stockholders were pleased. In October of 1980 the stock price hit an all-time high of just under $60 a share. But, despite the year's record sales, the quarter-to-quarter growth was minimal, with the $165 million in shipments split roughly into four $40 million chunks. Earnings per share actually declined in the third quarter and slipped further in the fourth. Of course, the growth rate of the past five years was unsustainable; everyone knew that. But the world's electronics manufacturing plants were filling up with thousands of Teradyne test systems, and this growing base, along with Teradyne's reputation for quality and its hefty engineering budget, suggested that the best was yet to come.

Good as the test system business was, the backplane business was even better, as sales at the Connection Systems Division rolled from $13.6 million in 1978 to $22.8 million in 1979 to $33.3 million in 1980. Military contractors still accounted for more than half the Division's sales, but commercial sales were growing fast, and Teradyne Connection Systems (TCS) was about to engineer the kind of military-to-commercial transition that other companies dream about but rarely pull off.

Despite expansions in Boston, Woodland Hills, and Nashua, the Company was out of space again, and once again Teradyne began searching for underused property near the Essex and Lincoln Street plants. The search led to a boxy, 10-story building on Harrison Avenue, just a few blocks away from Essex Street. The pile had been built at the turn of the century as a drug company warehouse. Its major tenant was a mailing service known as "Hub Mail," and a large sign on the building's roof proclaimed that fact. Other tenants included a tie maker (Esquire Neckwear), a clothing manufacturer (Pendleton), and various other garment producers.

The building was ugly in the extreme, and elevated trains ran by one side, almost close enough to touch. It would "need work," to put it mildly, but it did have 228,000 square feet of space, such as it was, and the reinforced concrete skeleton was sturdy enough. In any event, Owen Robbins saw the possibilities and commenced negotiations with the building's owners, the brothers Weisberg. For tax reasons, they were unwilling to sell on the spot, so a 25-year lease was executed on January 1, 1980, for $250,000 a year for the first five years and a total of $7.8 million thereafter, with an option to buy it for $3.2 million becoming effective in 1993.

The plan for the building was ambitious: It would be totally gutted inside and given a new skin on the outside, to become an appropriately handsome headquarters for the world's largest maker of ATE. (Within sight of Teradyne, a sign proclaimed that the Gillette plant was "World Shaving Headquarters." Harrison Avenue would be, plausibly, "World Testing Headquarters.") In time, Teradyne would also acquire adjacent properties to permit the construction of an attached garage. Robbins's vision for Harrison Avenue was uncannily on target. The new skin was installed (not without missteps), the entire interior was redone, all services were updated, an 11th-floor skylit cafeteria was added, and Boston soon had a most attractive addition to its skyline, where an eyesore had stood before. It was not perfect — secretaries transferred from Essex Street grumbled that they could no longer walk to Jordan Marsh or Filene's, and shuttle buses would be needed to ferry employees between buildings and parking lot — but once again Teradyne had scored a real estate coup.

Two more Vice Presidents were appointed in 1980: One was George d'Arbeloff, who was managing the Company's field service operations. The other was the author of this book, responsible for the Company's relations with the press, Wall Street, and its various other publics.

PART FOUR

The 1980s and the Threat from Japan

16

On Top of the World

In 1980 the semiconductor industry was for all practical purposes the private preserve of U.S.-based companies. The two largest chip makers were Texas Instruments and Motorola, and National Semiconductor, Intel, and Fairchild were also among the top 10.[1] The semiconductor equipment industry was also U.S.-dominated, with the first nine places all going to domestic firms. They were, in order of size: Perkin Elmer, GCA, Applied Materials, Fairchild (Sentry), Varian, Teradyne, Eaton, General Signal, and Kulicke & Soffa. Barely making it into the top 10 was the Japanese company Takeda Riken.

Ten years later, the world of semiconductor equipment would have turned upside down. The two giants, Perkin Elmer and GCA, would be gone from the top 10 list. So would Eaton and Kulicke & Soffa. Four of the five leaders in 1990 would be Japanese companies. Teradyne would have dropped to number 9. The situation mirrored that in semiconductors, where the Japanese would claim four of the top five positions. Worse, the prevailing view at the end of the 1980s was that the

[1]These rankings, along with those that follow, are per VLSI Research. The inclusion of IBM, a captive supplier usually omitted from such rankings, would have made the U.S. appear even more dominant in 1980.

contest was over, that the United States had lost the chip war just as surely as it had lost the automobile war.

Yet Teradyne at the start of the decade looked unstoppable. Sales had galloped from $38 million in 1975 to $165 million in 1980. In every one of its major markets — digital IC test, linear IC test, memory test, board test, and laser trimming, the Company held the number one or number two position. Its Connection Systems Division boasted a long string of growth years, and in the 4TEL subscriber-line test system the Company had a product of almost unimaginable potential.

Management seemed to be doing everything right: It had invested early in Europe and Japan, and those investments were paying off, with 40 percent of sales coming from overseas. Investors were making money on their Teradyne stock, and this group included most of the Company's 3000 employees. The Company had survived a major shakeup in the mid-70s, when one key manager left to form a competing company and another was cashiered over a policy disagreement, and it had come roaring back from adversity.

The management line-up was easily the strongest in the ATE industry. Co-founder and CEO Alex d'Arbeloff had surrounded himself with a hand-picked team that worked effectively with him and with each other. It was clear to anyone observing the day-to-day operations of the Company that two executives, Owen Robbins (CFO and de facto COO) and Jim Prestridge (running the all-important digital IC test business), ranked just below d'Arbeloff in power and influence, but this was not something to be acknowledged openly. In fact, when a visiting securities analyst asked, "Who is number 2?" Robbins immediately snapped, "There *is* no number 2." D'Arbeloff would have surely given the same answer, not only because of his habit of tap dancing around such issues. Joe Lassiter, a young ex-MIT professor recruited by d'Arbeloff in the mid-70s, had blazed his own trail as manager of the Company's board test business, was well liked, and was clearly one of the Company's leaders, if anyone bothered to think about succession.

But succession was the last thing on d'Arbeloff's mind in 1980. After some rocky years wrestling with the cyclicality of the business and the replacement of some key managers, he was now in top form. He had the loyalty and respect of his management team, he chaired a Board of Directors whose support he could count on, he kept in excellent physical shape, and he immensely enjoyed what he was doing. Home in Brookline, he lived with his wife Brit, two sons, and two daughters. His brothers Dee and George were also nearby, the former the CEO of Millipore, the latter a Teradyne vice president. The family, in the d'Arbeloff tradition, was close-knit.

At the 1980 Semicon trade show in San Mateo, California, the semiconductor equipment industry was feeling invincible. Show attendees marveled at the mammoth booths of front-end equipment makers like GCA and Perkin-Elmer, both growing even faster than Teradyne, and the word on every securities analyst's lips was "pervasiveness." Semiconductors, so the story went, would soon be used in automobiles, telephones, and virtually every kind of industrial machinery. The sky was the limit.

Some Semicon attendees were showing off a new Sony product called Walkman. Just introduced into the United States, the portable tape player impressed all those who heard its brilliant stereophonic output, and some were comforted by the thought that a Japan preoccupied with consumer gadgets might be less of a threat in the industrial arena. Canon, Takeda-Riken, and other Japanese companies were trying to become world players in the semiconductor equipment business, but their efforts were so far meeting with limited success. The Directors of SEMI, the industry trade group sponsoring Semicon, huddled over the issue of whether to grant Japanese companies full membership and exhibiting privileges. Some argued that the organization, whose name stood for Semiconductor and Equipment Materials Institute, was founded as a U.S. group and should stay that way, but most members of the SEMI Board favored a wide-open organization. Eventually, the majority ruled. Henceforth, SEMI would stand for Semiconductor Equipment and Materials *International.*

After the dizzying pace of the previous four years, a downturn was overdue, and Teradyne's sales in 1981 slipped back by about 3 percent — hardly the kind of blow that Teradyne had taken in 1971 and 1975. Moreover, the Company managed to earn money for the year — not a great deal of money, but $5 million pre-tax, even in the face of a 22-percent boost in engineering expenses. The necessity of funding key development programs, no matter what the current state of business, was an article of Teradyne faith from the very beginning. It was a matter of believing in the longer-term viability of the business. Fortified by this conviction, Teradyne always saw downturns as aberrations, pauses in the long climb to greatness. They were also seen as opportunities to gain market share from weaker competitors.

A tangible demonstration of Teradyne's faith in its future was the fleet of bulldozers clearing land in Deerfield, Illinois, the site of a new plant for Teradyne Central, and in Nashua, New Hampshire, where Connection Systems, having built and occupied its new headquarters, was again expanding. In

Boston, workers began turning the ugly duckling on Harrison Avenue into a swan (or at least a presentable goose), and the Digital Systems plant in Woodland Hills, California was being expanded.

The officer corps was also expanded in 1981, with the appointment of George Chamillard as Vice President in charge of ATE Operations, and the Board of Directors gained a seventh member with the election of John McArthur, Dean of Faculty at the Harvard Business School.

Although quarterly sales were stalled on a $40 million plateau throughout 1981, there was enough positive news on the product front to keep confidence high. The analog LSI and memory testers were selling well, and the L200 "combinational" board test system was proving once again the market's willingness to pay up for extraordinary performance. A $30 million 4TEL order from GTE was proof positive, if more was needed, that Teradyne had crashed the telephone club. To emphasize the point, Teradyne Central introduced a new product, called SCOT, to test the old tried-and-true stepper switches that were still used in vast numbers by telephone operating companies.

The most important of all the 1981 product announcements was the long-awaited debut of Teradyne's first VLSI (very large-scale integration) test system, the J941. D'Arbeloff and Prestridge had anguished over Teradyne's failure to mount a competitive challenge to Fairchild's Sentry test system and had considered entering the game via acquisition. Then engineering came to the rescue, when digital systems designer Gary Gillette, noting the steep slide in memory prices, concocted a memory-intensive architecture that enabled a system to deliver long test patterns at high speed — an innovation that increased test speeds five- or ten-fold. It was just the opening d'Arbeloff and Prestridge were looking for, and Gillette was given his marching orders and his budget. In 1981 the system was ready, and, even though the Sentry still held the Intel account, there were other makers of microprocessors to chase. Also, Intel itself was readying its next-generation microprocessor, the 80286, and this business appeared to be up for grabs.

As if making up for lost time, Teradyne announced that the J941 would be just one of *three* new VLSI test systems en route to the market. There would later be a higher-performance system, the J967, and a lower-cost version, the J983. (Students of Teradyne numerology would note that all three system designators were prime numbers.)

The J941, like its immediate predecessors, used Teradyne's own M365 computer as controller. The M365 had been a workhorse ever since DeWolf had developed it a decade earlier, and by any measure it was a fine computer, Teradyne-rugged and fast enough to do what it had to do. But by 1981 the computer industry had matured to the point where Teradyne could no longer defend its use of a home-brew computer by citing the lack of a commercial alternative, and the Company decided that the time had come to let someone else add this particular value to its test systems. So a small task force was assembled to make the all-important choice of computer for Teradyne's next generations of IC test systems. The L200 "supertester" from the board test group was VAX-based, and there were strong DEC advocates in the Company, but the semiconductor test groups felt that DEC had fallen behind technically and that their best solution lay elsewhere.

The task force quickly narrowed the field down to four contenders: Digital Equipment Corporation, then supplying Teradyne not only the L200's VAX but PDP-11s for networking applications; a small Massachusetts start-up named MassComp; Apollo Computer (later acquired by Hewlett-Packard); and Sun Microsystems, a small but promising company in the Bay Area. The decision, which would turn out to be providential, was to standardize on the Sun work station and UNIX operating software. Sun workstations would control the two VLSI test systems being developed as companions to the J941, along with the next generation of analog test systems.

The J941 was a well-designed system, with a number of features that could not be matched by the legacy-bound Sentry. For instance, in implementing Gillette's proprietary pattern generator, Teradyne stored 264,000 bits of memory behind each of its 96 pins, or channels, so that users could run long test patterns without reloading the local memory. Competing systems typically ran 4000 or 8000 bits at a time, forcing test programmers to break test patterns into manageable bursts.

The new system was also wrapped in a handsome new package, created by Paul d'Entremont, a talented mechanical designer hired from GenRad. Thus the rugged old kiosk slipped into history, although Teradyne kiosk-mounted systems would survive on factory floors throughout the world for a great many years. The d'Entremont package, featuring cream-colored panels with Teradyne-green accents, would prove even more durable, still looking contemporary 20 years after its debut.

The J941 quickly became a prime contender for the Intel business. The chief competitor was a system from GenRad (formerly General Radio), a leader in the board test business but a newcomer to semiconductor testing. Massachusetts-based GenRad had recently entered the business by hiring a team of Fairchild defectors and setting them up in the Bay Area as

GenRad STI. The GenRad system was credible, but the J941, with its unique pattern generator, was much more than that, and Teradyne could moreover point to its unquestioned dedication to the semiconductor test business and to its greater engineering resources. In a decision that effectively knocked GenRad out of the business, the J941 was declared the system of choice, not only for Intel's 80286 microprocessors, but inevitably for the "286" chips made by the many companies licensed by Intel, including Advanced Micro Devices, Siemens, and Philips. (Intel would come to rethink its policy of second-sourcing its microprocessors, and the J941 would be the last test system to get a free ride into other accounts on Intel's coat-tails.) In time, about 150 J941 systems would be shipped, and years after Teradyne discontinued production it was still in demand on the used-equipment market.

Meanwhile, the J941's predecessor, the J325, refused to die, and for an extended period Teradyne was to have the best of both worlds, with two generations of logic test systems simultaneously generating significant revenue. The J941's companion systems, the J983 and the J967, were less successful, but they did establish the "full-line" concept.

In memory test, Teradyne again went the three-system route, offering the high-end J389, the mid-range (and best selling) J387, and the low-end J385. Thus here and elsewhere Teradyne was in effect offering the good-better-best choices of a Sears Roebuck catalogue. The memory testers, along with the J941, shared a common high-level programming language called "Pascal-T," designed to simplify test program writing. The difficulties of writing test programs had long been the bane of the industry, and Teradyne made heroic attempts to cut the chore down to size, but there was always a Catch-22:

The customer spectrum included at one end programming virtuosos who gladly accepted a difficult language in exchange for the added flexibility it gave them, and at the other end programmers who wanted things made easy, and damn the flexibility. You could not please both types, though Teradyne tried to effect a reasonable compromise.

In the Industrial/Consumer Division, headed by John Turner, the A300 Analog LSI Test System once again stamped Teradyne as the team to beat in linear IC testing. With much stronger digital capability than its predecessor, the A300 also began to stake Teradyne's claim in the nascent mixed-signal test market, where LTX held a dominant position. The A300 line was sold in several market-oriented formats: the A310 series for analog LSI ICs such as operational amplifiers, codecs, voltage regulators, etc.; the A320 series, which came with a laser trimmer to adjust hybrid circuits; the A330 series, which included a more precise laser trimmer for use repairing redundant memories; and the low-end A351 for linear ICs. By the end of 1981, the A300 line was a best seller, with over 200 units shipped.

The Manufacturing Systems Division, spurred by the market's love affair with the L200, expanded its line of network continuity testers to cover unloaded and loaded boards as well as backplanes. The sales force was still trying to figure out how to sell the L621 board test system to engineers, and the Division was aggressively marketing the LASAR software as part of a VAX-based test pattern generator.

The Company, then, ran with five distinct product groups as the decade of the 80s began. In Boston were the Industrial/Consumer Division, headed by Turner, and Lassiter's Manufacturing Systems Division. In Woodland Hills,

California, Prestridge oversaw the digital IC test (logic and memory) businesses. Teradyne Connection Systems, still managed by founders Hunter and Johnson, and Teradyne Central, under Russ Ashdown, rounded out the line. Five good businesses, five good organizations headed by talented managers. There was every reason to believe that as soon as the recession ended Teradyne would take off again.

And take off it did, first quietly, with a 10-percent sales gain in 1982, then riotously, with sales almost doubling in the following three years, to $389 million. The burst was similar in magnitude to that of the late 1970s, but this time Teradyne was hitting on all cylinders. Profits as well as sales soared, as Robbins squeezed more and more efficiencies from operations. After-tax profit as a percentage of sales stood at an anemic 2.4 percent in 1982; three years later it was 11.1 percent.

Some of this was simply the magical effect of volume on margins. Teradyne is a "high fixed cost" operation, shouldering the high salaries of its engineers and other professionals in good times and bad. When sales drop, profits drop more. But as sales climb, more and more profit falls to the bottom line. The incremental margins in Teradyne's business — the share of each additional dollar of sales that turns into profit — tend to be very good. This is especially so if shipments consist largely of high-volume systems.

With two separate centers of semiconductor test activity, one in Boston under Turner, one in California under Prestridge, there was inevitably competition for resources. Of course, such competition is present in any multi-division company, and to a degree it is healthy. In this case, d'Arbeloff saw it as distracting, and he decided that it was time to reunite the Company's semiconductor test operations under a single leader. Prestridge was the obvious candidate for the job. Thus the Company formed the Semiconductor Test Group, embracing a digital (logic plus memory) division in California and a linear and mixed-signal division in Boston. Ed Rogas, now with six years of the semiconductor wars under his belt, was ready to take the helm in California, and he was appointed Manager of the Semiconductor Test Division (a misleading label, since the west-coast operation accounted for only half the Company's semiconductor test sales).

Shortly thereafter John Turner left the Company to enter the venture capital field, and Prestridge chose George d'Arbeloff to replace him as ICD Manager. Tom Newman succeeded d'Arbeloff as Field Service Manager.

In 1982, Teradyne Connection Systems moved into its gleaming new $10 million plant on Simon Street in Nashua, pulling the plug on its original plant in Lowell. The new plant boasted a highly unusual architecture, with two main manufacturing floors surrounded by a mezzanine of office space. The plant won several design awards, and first-time visitors were always taken with the unorthodox layout, the sodium lighting that bathed the factory floor in a peach-colored glow, and the rows of Teradyne-designed assembly and inspection machines that existed nowhere else in the world. The plant's woodsy setting, featuring jogging trails along the bank of a historic old canal, added to its appeal to present and prospective employees.

Just a mile or so away from the new TCS headquarters, the Division's first Nashua plant, on Pittsburgh Avenue, was converted into a printed-circuit facility, a major undertaking requiring not only the installation of expensive photographic, fabrication, and plating equipment, but massive waste treatment systems to meet environmental standards.

The investment in a PC shop was seen as a strategic, if costly, necessity. The wire-wrapped backplane was starting to give way to the PC "motherboard," and the trend was clearly irreversible. At some point, TCS's future would depend on its ability to produce the large, multi-layer boards its customers would demand.

There was one problem: Demand for printed-circuit backplanes had not yet reached the level required to support the large investment Teradyne was making at Pittsburgh Avenue. The solution: Use the new facility to produce the large volumes of daughter boards needed by Teradyne's ATE divisions. Then, when the printed-circuit backplane business reached critical mass, the daughter boards would be off-loaded to subcontractors, and Pittsburgh Avenue would begin doing the job it was built to do.

The maneuver worked as planned. Pittsburgh Avenue ground out daughter boards for Teradyne's ATE divisions (and, for good measure, for Wang Laboratories) for a few years, then became wholly dedicated to the production of the very large, very thick backplane boards that only a handful of plants in the world could handle.

In Deerfield, Illinois, Teradyne Central was building its own new 65,000-square-foot plant on a 13-acre site. This plant, too, featured an innovative design, with sharp angles and a glass and steel exterior giving it a rakishly modern look. Deerfield was near the Division's old site in Northbrook, so commuting patterns were not unduly disturbed, and it was a quick drive from the new plant to O'Hare. Teradyne people, like most high-tech people, tended to spend huge chunks of their lives on airplanes.

And work quickened at Harrison Avenue in downtown Boston, where the unsightliness of the old Hub Mail Building was disappearing under a beige veneer of a cementitious material sold under the trade name "STO."

The stock market reacted predictably to the good news coming from the Company throughout 1982. The stock, hovering around 30 in August, climbed to the mid 40s in October, split two-for-one, then kept rising, into the 30s. Investment bankers converged on the Essex Street offices, trying to convince d'Arbeloff and Robbins that the "circus was in town" and that it was time to "do a deal." Finally, the Company agreed to sell $50 million worth of convertible debentures. Carrying a 9 percent coupon and with a conversion price less than 20 percent above the market, they were seen for the good deal they were and quickly bought up.

The Company's good fortune notwithstanding, d'Arbeloff was nervously eyeing the signs of a threat on the western horizon. Japan's Takeda-Riken had introduced a line of memory testers and in 1982 began selling these in the United States, under the trade name "Advantest." Teradyne's memory test line was contributing heavily to its top and bottom lines (bookings were $30 million in 1982, $45 million in 1983), but almost all memory test sales were to U.S. customers. Meanwhile, Japanese semiconductor makers had begun a full-scale assault on the world memory market, and the clouds were forming over U.S. producers like Intel (which had created the memory market in 1971), Mostek (the dominant supplier of the 4K DRAM), Texas Instruments, and AT&T.

An earthquake of sorts had jolted the memory market two years earlier, when Hewlett-Packard revealed at a technical conference that the 16K DRAMs it was receiving from its U.S. suppliers were of distinctly lower quality than those received from its Japanese suppliers. The initial U.S. reaction was denial, but H-P's credibility was beyond question, and by 1982 Intel's conference rooms were alive with debates over whether that Company should stay in the memory business. It was not an easy call. On the one hand, argued Intel's Gordon Moore and others, the huge volumes and the iterative nature of memories made these devices ideal platforms on which to hone new fabrication processes. On the other hand, the major Japanese memory producers — NEC, Fujitsu, Hitachi, Toshiba — were all giant, multi-product manufacturers that could easily absorb losses in semiconductors in order to achieve larger goals — a luxury denied most U.S. chip makers.

U.S. semiconductor executives made pilgrimages to Washington, asking their government to retaliate against the Japanese for "dumping" and for other allegedly anti-competitive acts, but the handwriting was on the wall when Fujitsu and Hitachi introduced the world's first 256K DRAMs in 1982, demonstrating a clear technology lead over their U.S. competitors.

Watching all this play out, d'Arbeloff knew that Teradyne could not afford to lie back and await the Japanese onslaught, but would have to raise the stakes in Japan, expanding Teradyne's presence there to include engineering and manufacturing as well as sales and service. In other words, Teradyne would have to build a credible Japanese company, and that meant, first of all, finding and recruiting the necessary leadership.

Characteristically, he sought expert advice, in the person of Harvard Professor Ezra Vogel, an authority on Japan and the author of the best-selling "Japan as Number One." Could Professor Vogel help Teradyne find the talent necessary to build a stronger presence in Japan? The answer was yes. By a stroke of good fortune, Vogel had a perfect candidate: a young Harvard graduate who was fluent in Japanese. His name was Rick Dyck, and Vogel was the advisor for his Ph.D. thesis, whose subject was, of all things, the Japanese semiconductor industry.

Dyck knew little about electronics, but this was immaterial to d'Arbeloff, who had already turned an oceanography professor named Lassiter into a credible ATE expert. Dyck was hired in 1982, and a tutor was signed to teach him the basics of electronics. Within a year, Dyck was installed at Teradyne's wobbly Japan operation, whose annual sales had just plummeted from $20 million to $12 million.

Dyck was the fourth Teradyne manager in Tokyo. The first, Stu Peters, had laid a solid foundation in Japan and was revered by Teradyne's Japanese staff. He was now running Teradyne's European organization out of Weybridge, England. Peters was succeeded in Japan by Chris Davison, an English engineer transferred from Teradyne's European operations, and he in turn was succeeded by a Japanese sales executive named Hiroya Yano. Of the four, only Dyck was equipped by background and education to communicate comfortably and fluently with both his Teradyne colleagues in the U.S. and his Japanese employees and customers. It was to prove a winning combination.

The charter d'Arbeloff handed Dyck was to build a manufacturing and design presence in Japan to complement its sales and support activities. The memory situation was especially troubling to d'Arbeloff, who saw that Teradyne had erred grievously in designing its memory test systems around the needs of its U.S. customers, ignoring inputs from Japan. Thus, while Japanese tester makers Takeda-Riken and Ando were designing systems that could test large numbers of memory chips in parallel, Teradyne's memory test engineers, tuned to the U.S. market, were sure their system's superior ability to repair redundant memories would carry the day. Parallel testing, they insisted, was a non-issue. But to the Japanese, watching DRAM prices fall 70 percent in a single year, parallel testing was *the* issue.

In 1982, as Dyck landed in Tokyo, all this was more a nuisance than a crisis, for Teradyne's sales were about to explode. A modest 10-percent gain in 1982 was followed by a 42-percent increase in 1983. Memory makers such as Mostek ordered Teradyne testers by the dozens, many of these paired with Teradyne laser trim systems for repairing redundant memories. The new linear tester, the A360, was enormously successful in the U.S. and, more significantly, it scored a direct hit in the Japanese consumer electronics market. In fact, equipped with appropriate video and RF options, it became the Japanese (and thus the world) standard for testing TV and VCR devices. Dyck thus had his first real breakthrough and a foundation on which to build the larger Japanese presence d'Arbeloff had ordered.

The stock market again responded to Teradyne's good fortune, the stock climbing from 28 at the start of the year to more than 60 in July, when another two-for-one split took effect. The bonds that were sold in 1982 were converted into common stock at the earliest possible date, in October, 1983. In the fourth quarter of 1983, the Company passed a new milepost, booking more than $100 million worth of orders. For the full year, sales at the Semiconductor Test Division in California (the digital IC test operation) doubled, and the search was on again for expansion space.

Thanks largely to Intel and its 80286 licensees, the J941 VLSI Test System was able to post a profit in its first full year in production. In memory test, the Company was the hands-down market leader in the United States, and the spurt in memory tester sales was accompanied by strong sales of Teradyne laser trim systems used to repair memory chips.

The laser technology was impressive. The laser spot was positioned on the memory, quickly and accurately, by a pair of galvanometer-controlled mirrors, which pivoted in jeweled bearings. Ingenious software calibrated the laser's position, verified the repairability of the memory, and figured out how to perform the repair most quickly. In those days, long before laser scanners, CDs, and laser disks became commonplace, Teradyne was one of only a handful of companies using lasers to do real work, for real money.

The A360 hit a system-a-day stride in the fourth quarter, and a new technical center in Tokyo helped boost tester sales in that market. The L200 Circuit Board Test System, the target of so much skepticism when it was introduced two years before, was now a huge winner, with more than 100 systems sold — at an average price of about a million dollars — by the end of 1983. The Manufacturing Systems Division was also seeing a growing demand for its LASAR test generation software.

The sales records were falling at Teradyne Central, too. The 4TEL system, now a standard in the U.S. independent telephone network, established a beachhead in Europe with orders from the Hull (UK) telephone company. These were followed by orders from British Telecom and Belgium's telephone system, leading the Division to announce plans for a sales and engineering center near London. And at Teradyne Connection Systems, whose commercial business was now roughly half the total, record sales prompted the building of a 100,000-square-foot addition to the Nashua plant it had opened only a year or so before.

The Company's long run as a maker of transistor and diode testers came to a quiet end in 1983. The line no longer produced the kind of sales growth the new Teradyne required. Over a span of more than two decades, Teradyne had shipped thousands of diode and transistor testers to plants throughout the world, and many long-time Teradyne customers were no doubt grieved by Teradyne's exit from the business. To ease the pain, the Company took extraordinary steps to support the discontinued systems. In fact, working copies of most of Teradyne's classic old testers have been maintained for many years in the Boston and California plants, expressly for the purpose of back-stopping the customer base.

As sales surged in the early 80s, profits grew even faster. Owen Robbins, now steering operations as well as finance with a sure hand, began a crusade to improve the Company's asset utilization. As he saw it, a huge payoff was there for the asking if the Company could reduce the asset base required to support each dollar's worth of sales. Robbins drove down working capital relentlessly, streamlining operations and hectoring customers for faster payment of receivables. The latter necessarily involved sales and service no less than finance, since customers could be expected to resist pressures for payment as long as there was an excuse (e.g., an incomplete or inoperative system) to procrastinate. It was essential, then, that they not be given this excuse.

The campaign to improve asset utilization paid off handsomely: At year-end 1982, assets amounted to 94.2 percent of sales, annualized at the fourth-quarter rate. One year later, the figure had been cut to 73.6 percent. Tens of millions of dollars had thus been created out of thin air, just as surely as if Owen Robbins had found it on the sidewalk at Harrison Avenue. In

1983, it was gravy; by the end of the 1980s, it was to play a large part in Teradyne's rise from the ashes.

Head count at the end of 1983 totaled 3900, up 700 over the year-earlier total. Of these, a large number were employed in the field support of the Company's rapidly growing installed base of test systems. The test systems were complex, and they were vital to their owners' operations, so any sizable ATE manufacturer was obliged to field an army of service technicians. But that was just for openers. Customers looked to Teradyne, not just for system maintenance and repair, but for help in using the systems more effectively. Teradyne had always provided such help, but now the effort was formalized with the creation of the Technical Applications Group (TAG), which became a separate, worldwide corps of hundreds of engineers. At times these specialists might be called on to help close a sale — perhaps by writing a program for a key part — and at other times they might produce hardware or software options for a valued customer.

Some of the options devised by the TAG engineers were added to the catalog and became important revenue producers in their own right. Thus, if the design engineers were the composers of the Company's new products, it was the applications engineers who arranged the orchestrations. To name some of them is to risk slighting others, but among Teradyne's hundreds of applications specialists were some whose contributions are legendary: Ken Neves, John Donaldson, and Al Tajeda for their work in memory testing, Cristo DaCosta and Eric Larson for their innovations in VLSI logic test, Yves Destenaves and Joe Serra for their brilliant work in mixed-signal testing, and Hock Chiang, whose dedication and competence made him a favorite of Teradyne's Asian customers.

Staying close to the customer was a central tenet of d'Arbeloff's regime, as he realized that, with proper care and feeding, many of these customers would in effect become extensions of the Teradyne sales force as well as a source of valuable guidance in the charting of product road maps. A concrete expression of this philosophy was the Teradyne Users' Group, called TUG. TUG served to unite Teradyne's key customers, offering them, through newsletters and conferences, the opportunity to engage other customers as well as Teradyne engineers in discussions of test techniques. Nominally at least, it was the customers who ran TUG, the customers who presented papers, and the customers who set the agenda, while Teradyne simply provided the platform. In fact, TUG was a brilliantly executed marketing device as well as a boon to customers. Eventually, the annual TUG meeting, which would take place at various sites in the U.S. and abroad, would attract hundreds of customers, all gathered in common cause under the Teradyne umbrella.

Another core strength was a sales force that functioned smoothly and seamlessly across time zones and national boundaries. The major Teradyne customers were of course all multi-national companies, and Teradyne made sure that their plants in the United States, Europe, and Asia were all served by a single, coordinated sales force, working under common sales and service policies. The salesman who covered Motorola in Phoenix was expected to communicate regularly with the salesmen covering Motorola Toulouse, Motorola Austin, and Motorola everywhere else. And if a customer wanted to shift some of its Teradyne systems from a plant in Texas to a plant in Malaysia, it could be sure of the same quality of support at both ends of the journey. In fact, Teradyne in many instances took responsibility for the end-to-end transaction, crating the system, shipping it, and installing it at its new destination.

Like most companies, Teradyne also staged fairly elaborate week-long annual sales meetings, which might be annual or not, depending on the state of business and the available budget. An interval of eighteen months was typical, with the site chosen generally a resort where some time (a half day, typically) could be allotted to recreation. Among the resorts chosen for this ritual were Rancho Bernardo, near San Diego; Hamilton, Bermuda; Hilton Head, South Carolina (where the meeting began on the day of the 1987 stock-market crash); Intervale, New Hampshire; San Diego's Hotel Del Coronado; and, memorably, at the Carefree desert resort north of Phoenix, where Teradyne's Japanese and European salesmen delighted in reliving the Hollywood westerns they had seen as children. The sales meetings, which would later become "customer team meetings" to include applications and service engineers, were expensive investments, mostly because of the heavy travel expenses, but they were invaluable, both for their substance and for their morale-building.

After growing 42 percent in 1983, sales might have been expected to take a breather in 1984. Instead, they grew an astounding 55 percent, to $389 million. Return on average equity passed the 20-percent mark for the first time in many years. Common sense said that sales growth of 42 and 55 percent in successive years was not the natural order of things, that a reversion to the mean was inevitable. Some pointed to the pervasiveness of semiconductors and predicted a long, uninterrupted period of growth. But the stock market, increasingly uneasy about the ability of U.S. chip makers to compete against Japan, was not convinced. Although Teradyne's quarterly earnings grew from 39 cents a share in the first quarter to 67 cents four quarters later, the stock price languished in the 20s for most of the year.

The strong sales growth of the early 1980s propelled Teradyne to a point where it could claim without challenge to be the world's largest maker of ATE. Schlumberger's Sentry Division (the old Fairchild) was still a factor in logic test, and LTX was doing well in the mixed-signal market. GenRad was sputtering in semiconductor test but was still the market leader in board test. The Japanese ATE manufacturers, for the time being, were linking their fortunes to the Japanese semiconductor companies who were rising fast in the memory business. In the laser-trim business, Electro-Scientific Industries (ESI) was a strong competitor, and several small companies competed effectively in the backplane test and phone-line test markets. So the competitive scene was as cluttered as ever, with an important difference: One company, and only one, offered product lines spanning essentially the entire gamut of automatic testing, from microprocessors to memories to analog circuits to backplanes to boards to phone lines to laser trimming. The "broad-line" approach, on which Hewlett-Packard had built its instrument empire 30 years before, was carrying Teradyne to new heights year after year.

Yet the synergies that might have been expected to fall out of the broad-line approach were frustratingly difficult to achieve. D'Arbeloff had created an array of semi-autonomous divisions that resisted integration, and there was at headquarters no centralized engineering function to force them together. In the early 1980s a Product Integration Program ("PIP") had been launched under John Turner to investigate ways to leverage technology across product lines. The computer task force that had hammered out the agreement to standardize on the Sun workstation was another, more successful effort. Jim Prestridge, as Semiconductor Test Group Manager, was able to coax a measure of cooperation out of his digital and analog divisions. Otherwise, each group marched to its own drummer. In plain fact, the degree of cross-fertilization probably equaled that of most other multi-division companies and may even have surpassed some, because of the tendency of Teradyne engineers and managers to migrate from one division to another. But in the last analysis the success of the Company's IC test business could not be grafted onto the board test business, nor could the theoretical relationship between the phone-line test and linear IC test businesses be exploited.

D'Arbeloff was not one to worry about such things. He was a strong believer in decentralization or, as he would put it, forcing those who designed and sold the products to operate as close as possible to their customers, without any intervening layers of management. He fully understood the price he was paying in creating a decentralized organization, but he felt that the greater value lay in giving each division its head, to the greatest possible extent. He had full confidence in the executives leading the various divisions, and if there was to be technology-sharing, they, and not some headquarters commissar, were the best ones to drive it. Given the roaring success of the Company in the first half of the 1980s, d'Arbeloff's management techniques could hardly be questioned.

Although engineering and marketing were decentralized, there was no question where the seat of operations lay. This was unequivocally Robbins country, and it was a big spread, extending into the far corners of the Company. Robbins wielded his growing influence at the quarterly results/forecast meetings he had initiated a decade earlier, meetings which had become Teradyne's planning metronome.

Attendance at these meetings grew to the point where twenty or thirty people would regularly convene to report and scrutinize the quarterly affairs of five or six definable businesses. This would have been regarded as bureaucratic overkill by the Teradyne of Nick DeWolf's day, when all meetings involving more than two or three people were seen as essentially nonproductive. Now they were seen as a positive, because they forced so many middle managers on stage to play their roles in the central drama of the Company.

The results/forecast meetings were now taking place in a tenth-floor conference room at Harrison Avenue, for in 1984 Teradyne abandoned Essex Street and completed its move to its new headquarters. The new site was just south of the Massachusetts Turnpike and just east of Boston's "Orange Line" elevated train tracks, soon to go underground. Chinatown lay just across the Turnpike to the north, and Boston's South End, tired but showing signs of gentrification, sprawled to the south. Parking was troublesome as always, but the Company leased some space nearby and drew up plans for a garage adjoining the plant.

Although the Company fielded ATE's broadest product line, there were still a few gaps. One of these was filled when the Company purchased Adar Corporation from Scientific-Atlanta in 1984. Adar, which had earlier been an independent, marginally successful producer of memory testers, had been purchased by Scientific-Atlanta a few years earlier, and it had not been a good fit. Its product line featured the MX-19 Tester for non-volatile memories (EPROMs and the like), an area of considerable promise. The Company, whose management team included the indomitable Abe Bluestone as well as former Teradyne sales manager Tony Martinelli, was reconstituted as Teradyne's Burlington (MA) Division, and Phil Phillips, Memory Test Product Manager, was brought east from California to head the new Teradyne unit and to launch the development of a new tester for the nonvolatile market.

It was not to be. As designed, the new system was too pricey for its market, and a cyclical downturn in 1985 put the final nail in the coffin. The Burlington Division was closed, Bluestone migrated back to Teradyne headquarters to take on a staff assignment for his old friend d'Arbeloff, Martinelli wound up as a sales manager for the Telecommunications Division, and most of the other Burlington employees found positions elsewhere in Teradyne.

If Adar was a misfire, Teradyne Connection Systems was a smash hit. In 1984, 16 years after its founding, TCS was running at about a $30 million annual rate — not all that different from what its parent Company had achieved at the same age. The products were innovative, and the custom-designed fabrication equipment gave the Division a competitive edge. And the profit margins were solid. But d'Arbeloff, ever restless, found causes for concern. The Division's commercial business was growing, but its commercial customers were smaller, second-tier manufacturers. There were almost no sales outside North America. And the management team that had led the Division so brilliantly in its early years had yet to recruit a successor group capable of driving the Division to its next level. (One of the founding trio, Sales Manager Tom Neavitt, had left in 1977 to form a company representing the TCS line in New England. Neavitt, a jolly, sales-savvy pro, would continue representing TCS until he retired in 1985.)

In the mid-80s, d'Arbeloff forced the issue, setting in motion a series of appointments that would completely overhaul the Division's management. Windy Hunter, the Division's Manager, was asked to accept an assignment to study the Japanese market, and he was replaced as Division head by Len Johnson, another TCS founder and the Division's chief engineer. Then, in 1984, George Chamillard, Teradyne's top ATE operations manager, was transferred to Connection Systems as Sales and Marketing Manager. The appointment of a manufacturing executive to a top marketing post might have raised eyebrows at other companies, but not at Teradyne, where such cross-functional moves were often used to round out the portfolios of managers on the rise.

Johnson's tenure as Division Manager would be brief — an engineer at heart, he found little enjoyment in administration — but during his 18 months at the helm he did manage to jump-start the Division's anemic overseas activities by recruiting one of Teradyne's most valuable executives.

In 1984 Dennis O'Connell, then in Boston heading Teradyne's world-wide sales force, told d'Arbeloff of his desire to return permanently to his native England. D'Arbeloff, bowing to his wish but intent on keeping O'Connell if at all possible, suggested that he continue to run non-U.S. sales from a base in London. O'Connell gamely gave it a try, shuttling from Europe to Japan for several months before giving it up as a bad bet. He thus became available, and Len Johnson and George Chamillard, looking for the key that would unlock the overseas markets, jumped at the chance to land him. O'Connell would preside over TCS's European business until his retirement in 1992 when, for the first time, Europe contributed 10 percent of the Division's bookings.

By 1986, the scenario in Nashua was playing out as d'Arbeloff had probably foreseen. Chamillard took over as General Manager only a year and a half after joining the Division. Johnson once again ran engineering, and Hunter returned from Japan to take over the all-important commercial side of the business — reporting to Chamillard. And Dennis O'Connell was laying the groundwork for a substantially greater TCS presence in Europe.

Taking over at Nashua, Chamillard began recruiting talent from elsewhere in Teradyne and from other companies, and a new management team was soon taking shape.

Within a few years both Johnson and Hunter both decided to call it a day — Hunter after a stint as the Division's first TQM Manager. All three of the TCS founders had much to be proud of: They had taken a then-novel idea — the backplane as the foundation of an important business — and created a large, fast-growing, highly profitable enterprise around the basic concept. The parallels with what DeWolf and d'Arbeloff had done were obvious, and the Division's best was yet to come.

Under Dennis O'Connell, a single sales force had sold Teradyne's entire ATE line, including both semiconductor testers and board testers. His move to Europe was the catalyst for splitting that sales force into two groups, one selling semiconductor testers, one selling board testers. Responsibility for the semiconductor test line was given to John Casey, who had joined the Company eight years earlier as a salesman in Texas. Pre-Teradyne, Casey had worked in sales for Faultfinders, a board test company acquired by Schlumberger. Board test sales were placed under Jeff Hotchkiss.

The splitting of the ATE sales force was an important and somewhat controversial development. Powerful arguments could be made, and were made, for maintaining a unified sales force, in which the best sales talent could be leveraged across the entire ATE product line. There was also some fear that the Company, already fragmented, would find it increasingly difficult to preserve a common culture if more interdivisional walls were erected. On the other hand, some specialization had already occurred; a Teradyne salesman based in Silicon Valley was not likely to spend much time selling board testers. In fact, board test manager Lassiter correctly saw that his best chance for claiming the full attention of the sales force was to make it (or part of it) his own, dedicated sales force. And so the ATE sales force was divided, not without trauma.

In Japan, Rick Dyck remained in charge of the Company's rapidly expanded activities, including the Japan Division, a design and manufacturing unit newly formed under veteran Teradyne engineer Bill Attridge. In Europe, the vacuum created by O'Connell's departure triggered a quest for a new senior manager. The post was soon filled by David Hodgson, a seasoned executive who had managed the Marconi division of the U.K.'s General Electric Corporation.

Five new vice presidents were appointed in 1984: Loren Eaton (Personnel), Windy Hunter and Len Johnson of Connection Systems, Tom Newman (Customer Service), and Ed Rogas (Digital Systems).

17

• • •

On the Light Side

A cardinal element of Teradyne's philosophy was the Company's obligation to share the fruits of its success with those who made it possible. This implied something beyond salaries, which were competitive but usually no more than that. Stock options were used widely and effectively, but the payoff from these was at the mercy of a fickle stock market. There was also a Savings Plan, funded partly by an employee's contributions, partly by the gains from the exercise of stock options owned by the Plan.

The missing ingredient was a profit-sharing plan, and such a plan was added to the benefit mix in 1984. The essentials were fairly standard: Twice a year, ten percent of the pre-tax profit for the just-completed six-month period would be divided among the employees, with the distribution pro-rated on the basis of salary. Thus, for a given half year, the payout would be announced as so much per thousand dollars of salary. Although profits would be erratic over the rest of the decade, eventually the profit-sharing plan would prove a windfall for the employees, often inflating their annual salaries by 15 to 20 percent.

Once the profit-sharing scheme was devised, attention turned to the manner in which the payout would be communicated to the work force. Video tape was the chosen medium, and it was decided that a tape would be made each quarter, so that d'Arbeloff might communicate more effectively with an employee population now numbered in the thousands and spread across three continents. The quarterly videotapes were to become an institution, which d'Arbeloff would use to convey news, to promote various initiatives, to comment on the business outlook, and to prod employees to move new products to market faster.

The early video tapes showed d'Arbeloff as rather stiff and uncomfortable, and a variety of formats — TelePrompTer, notes, cue cards, etc. — were tried by employee communications writer-turned-TV-director Andy Porter, in an effort to induce a more conversational style. The most promising was an interview format, in which I served up softball questions and did my best to put d'Arbeloff at ease. In time d'Arbeloff became a master of the art, and by 1987 he was confident enough of his television skills to agree to take part in a video tape parody of the quarterly productions, to be shown at an international sales meeting.

The resulting gag tape was a resounding success and inevitably was followed by other gag tapes produced for other sales meetings. Still, the total audience for these gag tapes was relatively small, and most employees were never given a glimpse of d'Arbeloff's great talent for self-parody — or of his razor-sharp sense of humor. On the contrary, he often came across as dour and off-putting. But the facade was misleading; just under the surface, there always lurked a keen appreciation for the inane and ridiculous. The following excerpt from one of them cannot reproduce the facial expressions and impeccable timing by which d'Arbeloff made fun of d'Arbeloff, but they will give the reader a taste of what we are talking about.

Van Veen:	What TQM issue are you going to talk about today, Alex?
d'Arbeloff:	"Market-In."
Van Veen:	Which means?
d'Arbeloff:	It means listening to your customers. Not necessarily agreeing with them, but listening to them. You've got to listen in this business.
Van Veen:	Is this primarily a sales issue or a marketing issue, Alex?
d'Arbeloff:	(straight at camera) And too many people at Teradyne don't listen. You've got to listen.
Van Veen:	Where do we stand in the whole "Market-In" issue, Alex?
d'Arbeloff:	(droning on) Because if you don't listen, you don't learn anything, and if you don't learn anything, where are you?
Van Veen:	Perhaps now we ought to discuss the success you think we're having in TQM, over-all.
d'Arbeloff:	Now I think we ought to discuss the success we're having in TQM, over-all.

This was d'Arbeloff at his most ingratiating, playing the part of a self-absorbed executive, and it hit close enough to home to be very funny.

18

• • •

Decline and Fall

As Teradyne celebrated its silver anniversary, the storm clouds that d'Arbeloff had seen gathering on the western horizon broke over Teradyne. Takeda-Riken, now a formidable competitor throughout the world, signaled its designs on the U.S. market by renaming itself Advantest. Intel's Andy Grove, staring at a sea of red ink, bit the bullet and announced that Intel would exit the memory business. As if on cue, Mostek canceled orders for $3 million worth of Teradyne memory testers. In total, memory test orders plunged from $48 to $22 million, linear test orders were halved, and logic test bookings fared no better.

The laser trim business, which accounted for $48 million worth of orders in 1984, faced a new threat: Not only was the memory business under siege from Japan, but the process of repairing redundant memories was changing. Instead of repairing memories on-line, a procedure that required roughly one laser trimmer per tester, chip makers were shifting to off-line repair, which meant that a single trimmer could serve many test systems. Thus two of Teradyne's major revenue streams were imperiled.

There were some bright — or relatively bright — spots. Board test orders were given a boost as the ambitious (1000-pin, 80-MHz) L290 joined the line and as the lower-priced L210 began shipping in quantity. A new mixed-signal tester attempted, with some success, to cut into LTX's wide lead in that market. In Japan, Rick Dyck was making definite progress, especially in the consumer electronics sector, where the local competition was weak. (There was supreme irony in this fact, if anyone cared to notice.) Teradyne KK and the Japan Division moved their 150 employees into a new seven-story headquarters in the Shibuya section of Tokyo, and sales and service centers elsewhere in Japan and in Korea, Hong Kong, Taiwan, and Singapore underscored Teradyne's serious intentions in the Far East.

Sales at Teradyne Connection Systems grew slightly, and the Telecommunications Division (formerly Teradyne Central) added to its 4TEL installed base in the U.S. and Europe, shipped its 1000th SCOT tester for stepper central offices, introduced a 4TEL derivative (XTEL) for the Bell system, and came up with a device, called DMARC, that enabled telephone companies to distinguish faults on its line from faults inside the customer's premises. At first blush, DMARC seemed to offer Teradyne an enormous opportunity, as tens of millions of such devices would be needed — in theory. In the end, DMARC was another in a series of dry holes that resulted from this Division's opportunistic sorties. Another was a product called "The Electronic Receptionist," a highly sophisticated voice-mail system that was probably ahead of its time but definitely outside of the Division's mainstream.

There were other new-product excursions, usually highly imaginative, but they foundered for two reasons: First, the switch-makers that dominated the telephone equipment business — companies like AT&T, Northern Telecom, Alcatel, and Ericcson — jealously guarded what they saw as their turf, adding needed capabilities to the switches and thereby squeezing out independent suppliers like Teradyne. Second, despite the enormous and richly deserved success achieved by 4TEL, Teradyne was not willing or able to organize the resources needed to market a broad line of telecommunications gear. The SCOT Stepper Central Office tester had done well enough, but the stepper switch was en route to extinction, and its growth prospects were nil.

Although the Company was making undeniable progress on several fronts, 1985 was a poor year, with sales down 13.6 percent and net income down more than 50 percent. It was tempting to believe that this was just another in the series of short downturns that seemed to be the industry's and Teradyne's lot. Down years all seemed to end in 1 or 5 — 1971, 1975, 1981, 1985 — and every time the Company had rebounded quickly and dramatically. But this time there was a difference. The world of ATE had changed forever. A new player was on the block. Its name was Japan, Inc.

The dreariness that engulfed Alex d'Arbeloff in 1985 was deepened by personal tragedy. His younger brother Dee, who had been by his side throughout his entire life and whose business achievements (as CEO of Millipore) mirrored Alex's, fell victim to cancer in mid-year.

In June, the Company received a special request from Director Dan Gregory. Vice President George Bush was looking for high-tech exposure and would be available for a plant visit in the Boston area on July 15. Gregory's Republican connections passed the word to him, and he called Teradyne, asking if the Company would be interested in hosting the Vice President.

D'Arbeloff had always scrupulously avoided anything that smacked of politicizing the Company, but there had been a plant tour years earlier by a gubernatorial candidate named Dukakis, and in 1985 Mr. Bush was not a candidate but the sitting Vice President. It was quickly decided that Teradyne would be delighted to have Mr. Bush spend some time at the Company's headquarters in Boston.[1]

In the days before the scheduled visit, security personnel, accompanied by sniffing dogs, scoured the Harrison Avenue plant for any possible security problems. There was minimal disruption to the Company's operations, but there was some, and d'Arbeloff began to grumble. Then, while the Bush motorcade was making its way from Logan Airport to Teradyne, there was a final sweep of the building, with secret service agents relaying messages to each other via walkie-talkies. At one point an agent, attempting to solder a security device too close to the sprinkler system, set off the alarm, triggering an evacuation of the plant. When a second alarm caused a second evacuation, d'Arbeloff erupted, while associates tried to calm him. "I don't give a damn who's coming!" he shouted, making sure the Secret Service was within earshot, "I'm trying to run a business!"

Then the Vice President's limousine slipped into the Harrison Avenue Receiving area, the Vice President was greeted warmly by a smiling d'Arbeloff and a few Company officers. All trooped up to the tenth-floor conference room for a chat, then a quick plant tour, and then to the cafeteria, where the Vice President spoke to a large gathering of employees. He congratulated

[1]A decade later, another sitting Vice President would ask Teradyne to provide a backdrop for a speech. This time the politics were unmistakable: With the New Hampshire primaries approaching, Vice President Gore's team sought a stage at the Teradyne Connection Systems plant, in Nashua, to showcase his interest in technology. The Company obliged.

Teradyne on its success and paid tribute to an employee who had fled Vietnam as one of the "boat people." (A great many Vietnamese worked at Teradyne, mostly in California.) Flanking Mr. Bush was a beaming Alex d'Arbeloff, while the chastened Secret Service agents stood just offstage. (An extra fillip of excitement was occasioned by the fact that on that very day President Reagan was "under the knife," albeit only for polyp removal.) Then the event was over, Mr. Bush was off to Kennebunkport, and everyone breathed easier.

D'Arbeloff always believed that the best antidote for a case of falling sales was a heavy dose of new products, the more exciting the better. Now, hope was focused on a new mixed-signal test system, the A500 series. The "A5" development had begun in 1983, under the leadership of Dave Sulman, who had produced the breakthrough L200 before transferring from board test to the Industrial/Consumer Division in 1982. Unlike other mixed-signal systems that were essentially souped-up analog or digital systems, the A500 was designed from the ground up as a true mixed-signal test system, with a "vector bus" architecture that carried digital-analog synchronization to a level far beyond that of its forerunner, the A370. The A500 was a very ambitious system, with high built-in risks, just as the L200 had been a high-risk system. In this case, recalls Sulman, the risks were justified by Teradyne's anemic share of the mixed-signal arena. "When you have only 15 percent of the market," he says now, "it's okay to take risks."

What were the risks? Beyond those normally associated with any major system project, there was one monumental problem: The Sun-based A500 would not be compatible with Teradyne's A300-series installed base. Although this base was modest in mixed-signal, it was huge in the straight linear, or analog, market.

To major A300 users like National Semiconductor and Sanyo, a non-compatible A500 could be taken as betrayal. Thus, well before the A500 was announced, ICD's engineering staff was split in two, with one part continuing to develop upgrades and options for the A300 line, the other working on the A500. This dual effort eventually created a hybrid: an A500 that incorporated a Teradyne M365 computer, to run A300 programs and placate the installed base. But National bolted anyway, to LTX, and Sanyo turned to YEW (Yokogawa Electric Works), allied with Hewlett-Packard.

The A500 was formally introduced in January, 1986, with no orders in hand. (This was unusual. The Company's practice had been to defer new-product announcements until at least a few orders had been gathered.) After five months of hard selling, there were still no orders. Then Analog Devices bought a system, then Texas Instruments, then Telebras, South America's largest telecommunications manufacturer. Then the momentum started to build, as the mixed-signal world began to see what Teradyne was offering. Just as Sulman's L200 had given the board-test world functional and in-circuit testing in a single mainframe, so Sulman's Vector Bus A500 was offering producers of "system silicon" the chance to test the functional integrity as well as the components of their circuits.

Once the tide started to run in Teradyne's favor, the Industrial/Consumer Division was ready to roll. Division Manager George d'Arbeloff had assembled a first-class team: In addition to Engineering Manager Dave Sulman, he had as Marketing Manager Mike Bradley, another Coopers & Lybrand alumnus (and another in Teradyne's growing roster of Harvard MBAs). Bradley had joined Teradyne in 1979 and served a tour as Controller in the Customer Service Group before redirecting his career into marketing. The Division's

Production Manager was Teradyne veteran Asa Siggens, a laconic, highly able Teradyne manufacturing veteran.

As it became apparent that they had a winner on their hands, the ICD team parlayed their considerable resources into a major effort to wrest market leadership from LTX. Soon the press was abuzz with talk of the A500, Teradyne's new "LTX killer." Few at Teradyne were willing to go that far, and predictably, LTX mounted a counteract, under the trade name "Synchromaster." But the A500 was off and running, and it was unstoppable. In the coming decade, the A500 would propel Teradyne's share of the mixed-signal test market from 15 percent to a commanding 40 percent, as the market shares of Teradyne and LTX were reversed.

One of the characteristics mandated by Teradyne's founders was total candor in dealing with the outside world. (DeWolf, of course, carried this principle to an extreme.) Thus the 1986 Annual Report made no attempt, in the President's letter, to whitewash the dismal facts:

> Most of Teradyne's 26 years have been very good, and many of them have been spectacular. A small number... have been bad.
>
> 1986 was definitely one of the bad ones.
>
> ... we've always believed in keeping score in dollars, and it's hard to put a bright face on a year in which net profits fell 99 percent.
>
> (*The Wall Street Journal*, impressed, ran a story about this Report.)

When sales were skyrocketing in the early 1980s, Teradyne was forced to take steps to add capacity, and in 1985 ground was broken for a complex of three 120,000-square-foot buildings in Agoura Hills, California.[1] After the sale of the Woodland Hills site, the net cost of the expansion worked out to about $27 million. The expansion, flying into the teeth of Murphy's Law, marked the top of the cycle, and much of the new space would lie dormant for years.

In Japan, the greater Teradyne presence that d'Arbeloff had ordered was now a fact. A design and manufacturing unit, called the Japan Division, had been established, and engineers at this unit collaborated with the Industrial/Consumer Division in Boston and with a very demanding customer (Sony) to create a test system for image-sensing devices. The Company also scored heavily in the Japanese consumer-electronics market with the A300 Analog Test System, but it had still failed to crack the Japanese memory test market, although it did manage to sell Mitsubishi an impressive number of memory test systems and laser trimmers.

In late 1986, the Connection Systems Division established its own beachhead in Japan, assigning the newly hired Peter Kelley as its Asia Manager. Kelley, a former lawyer, had come to Teradyne following the footprints of Rick Dyck, learning Japanese at Harvard under Ezra Vogel. Over the next five years, he would preside over a growing TCS presence in the Far East, setting up a small manufacturing group as well as a sales and service office.

[1]The proposed buildings were only two stories high, but the Agoura Hills town fathers worried that they might eclipse their fine view of the hills to the south, and approval was momentarily in doubt. The day was won when Teradyne gave the Town the chance to select the exterior color of the buildings. The sand-colored complex is today one of the more attractive commercial complexes along the Ventura Freeway.

In Europe, the ATE, Connection Systems, and Telecommunications Divisions moved into a new regional headquarters in Bracknell, near London. The semiconductor test business seemed to track its U.S. counterpart, and business was down. Board test held its own, thanks to the sales of some large L200s to several telecommunications manufacturers.

But the big story in Europe in 1986 was the receipt of the largest single order in the Company's history: $28 million dollars' worth of 4TEL systems for the telephone system of the Netherlands. With Belgium already in the fold, and British Telecom nibbling, the 4TEL ship was now secured by two sturdy anchors: the U.S. independent telephone market and Europe. But d'Arbeloff, viewing the glass as half empty, feared that, unless 4TEL could penetrate the Bell system or Japan's NTT, it would soon run out of major targets of opportunity.

The addition of Cincinnati Bell's Chairman and CEO, Dwight Hibbard, to the Board of Directors in 1986 stirred d'Arbeloff's hopes. Cincinnati Bell was one of a very few U.S. companies that had managed to build a good business at NTT (selling software systems), and d'Arbeloff reasonably expected to benefit at the very least from Hibbard's experience. His enthusiasm was not shared by Telecommunications Division Manager Russ Ashdown, who doubted that the Japanese giant would allow Teradyne access to its network. They might buy a system or two to learn the technology, he cautioned, but they would never open their phone system to Teradyne. He would be proved right, but then, d'Arbeloff was also right in insisting that the effort be made.

Hibbard was not the only new director added in 1986. John Mulroney, President of Rohm and Haas came aboard the same year. Director Walter Lingle, the retired Procter & Gamble executive, retired from the Teradyne Board in 1986, and in the following year, the "P&G seat" was filled by Edwin Artzt, Vice Chairman of the consumer products giant.

Thus Teradyne, still a relatively small company, fielded a Board of Directors that included the CEO of Cincinnati Bell, the President of Rohm and Haas, the Dean of Faculty at the Harvard Business School, the Vice Chairman of Procter & Gamble, and a former Chairman of the New York Stock Exchange, among other distinguished members.

Every couple of months this Board would convene in the 10th floor conference room at Harrison Avenue to help chart the Company's course. At these meetings, the out-of-town directors would rely on d'Arbeloff's able and dedicated secretary, Kit (for Kathleen) Rosenbauer, to monitor logistics and to maintain communications links with their home bases. Typical of the top-flight executive secretary that she was, she handled all these chores and more with dispatch and good humor, rising to the occasion when a scheduling problem (e.g., a fogged-in Logan Airport) threatened to unhinge some Board member.

Sales in 1986 were $306 million, and net profit was barely above break-even, thanks to a capital gain on the sale of the old Essex Street building. Head count was held flat at 4300 (down from 1984's 5000), but it was obvious that, without a quick order surge, a layoff was inevitable.

D'Arbeloff's bias against acquisitions had been put to the test several times in Teradyne's history. Competitors Fairchild and GenRad had been the subjects of merger talks, and a number of smaller companies had been "shopped" to Teradyne. Although some of the merger discussions were quite advanced, in each case d'Arbeloff was able to walk away from the dance without regret.

Now, seemingly unable to get the Company back on its growth track, he was ripe for new ideas. Among the more interesting to come along was the expansion of Teradyne's presence in the fast-growing market for electronic design automation (EDA) software. The strategic rationale was the assumed convergence of design and test technologies. If Teradyne, already a major power in testing, could establish a strong position in design software, it could link both ends of the product creation process. The acquisition wheels thus began to spin.

Two small California companies were available: CASE Technologies and AIDA Corporation. CASE was a five-year-old company specializing in schematic capture and design entry software, while three-year-old AIDA supplied simulation software to ASIC (applications-specific IC) manufacturers. And, while the acquisition juices were flowing, why not add a third company? Zehntel, a veteran board-test firm based in Walnut Creek, California, was struggling and a likely acquisition candidate. Zehntel could bring about $40 million of annual sales to the table and thus give Teradyne the additional mass it needed to challenge market leader GenRad.

It all sounded reasonable to the Teradyne Board, and by year-end all three companies were added to the portfolio. For Zehntel, a public company, Teradyne paid 1,460,000 shares of its stock. CASE cost it another 930,000 shares, AIDA 1,875,000 shares. Thus, based on Teradyne's stock price of about $32 a share when the acquisitions were announced, the Company paid about $136 million (about 15 percent of its equity) for the three companies. The total sales of the three companies in 1987 were $54 million, and all three were operating in the red. The market did not give the deals a resounding cheer: Within three months, Teradyne's stock price dropped from the thirties to the mid-teens.

It is easy in hindsight to second-guess any acquisition, but Teradyne was taking what seemed at the time to be an intelligent risk. Zehntel was a sound bet, at a reasonable price. Although it had suffered three straight years of losses, it had a good track record over a long history, a large installed base, and an excellent reputation. It also came with a generous supply of ATE-savvy talent, including, for example, marketer Craig Pynn, later to become Teradyne's director of corporate marketing. CASE and AIDA were speculations, pure and simple, steeped in a culture that was light years removed from the no-frills, nuts-and-bolts Teradyne that Nick and Alex had built, but potential keys to the design-to-test linkage d'Arbeloff sought.

Zehntel CEO Ron Borrelli became a Teradyne Vice President, and it was left to Jeff Hotchkiss to integrate the CAE operations. All three acquisitions were placed within Lassiter's Assembly Test/EDA Group, the new name for the expanded Manufacturing Systems Division.

In Europe, semiconductor test sales manager David Hodgson decided to leave Teradyne, and the search was on for his successor. Before long, d'Arbeloff had his man: the European sales manager for Fairchild/Schlumberger Semiconductor, Roberto Stefanelli. Stefanelli's preference was to be based in his native Italy, and thus Teradyne's European semiconductor test headquarters moved from London to Milan — just a few miles from one of the largest concentrations of Teradyne test systems anywhere in the world, at SGS-Thomson's huge Agrate plant. SGS had long been one of Teradyne's most loyal customers, and it was fitting that Teradyne's new European control center was moving into its neighborhood. Stefanelli stepped into his new role as if he was born for the job and for Teradyne.

In Japan, Michael Schmidt, a 10-year veteran of the Industrial/Consumer Division, was appointed manager of the Japan Division, then losing $2 million a quarter on sales of the same amount. Schmidt soon perceived that the Division's near-term salvation lay, not in a new tester designed from the ground up in Japan, but in an amalgam of A300 and A500 technologies, coupled with some highly focused local applications work. It was the right formula. Winning a shoot-out against Advantest for Sony's image-sensor test business, JD turned the corner and became profitable less than two years after Schmidt's arrival.

In 1987, excluding the acquisitions, Teradyne's sales grew from $306 million to $324 million, but the bottom line was red, with a total loss of $15 million excluding acquisitions, more than $21 million with the acquisitions factored in. Heavy charges were taken to cover merger expenses, layoffs, and reconfiguration of the VLSI test line. The Company's balance sheet was still strong, the product line was competitive, and the A500 much more than that. But the fact remained that the Company had just reported its first-ever back-to-back down years, it was losing money, one of its principal product lines (memory testers) was under siege by the Japanese, and another (laser trimmers) was threatened by a shift in the way memories were being repaired.

Teradyne's greatest strength in times of stress was the quality of its management. The CEO was resolutely confident in the Company's future, and Prestridge and Robbins shared that confidence. Immediately under them were a dozen or so veterans of earlier campaigns: Joe Lassiter, looking for the magic bullet that would catapult Teradyne to the top of the board-test market; Jeff Hotchkiss, trying to make sense out of the DATA-CASE-AIDA combination; John McCabe, another Coopers & Lybrand alumnus, who had a long-term lease on the Controller's job; Treasurer Stu Osattin, who always seemed to be arm-wrestling with contractors, architects, and lawyers; Alex's younger brother George and his ICD team, riding the A500 whirlwind; Ed Rogas, running the digital operations in California; John Casey, working days and nights to direct a sales force stretching from Orange County to Milan; VLSI Product Manager Wayne Ponik; Chamillard and his new management team at Connection Systems; and Russ Ashdown at the Telecommunications Division. Reporting to them were dozens of other managers and future managers. The Company's year-after-year mining of the college campuses had paid off handsomely, in the form of layers of talent at the ready. Yet one question would not go away:

> If Teradyne had so much talent, why was it losing so much money?

The easy answer was Japan. The U.S. memory producers were now in full rout. Intel had thrown in the towel, and Mostek had been acquired by United Technologies, which was in the process of shuttering it.[1] AT&T, another major user of Teradyne memory testers, was also wavering and would eventually fold its hand. The notion that Japan was unfairly attacking U.S. industry ran deep in the country at the time, and the same industry

[1]Mostek still owed Teradyne several million dollars after canceling orders in 1985, and UT had set up a reserve to cover Mostek's liabilities. Teradyne sued to recover, and did. This was the only time in Teradyne's history that it sued a customer — though technically, it was not UT but Mostek that was the customer.

executives who had always resisted government intrusion in their affairs now marched on Washington, begging for protection. To his and Teradyne's everlasting credit, d'Arbeloff would have none of this. In his view, the problem was Teradyne's, not Tokyo's or Washington's, and it was up to Teradyne to solve it.

One Washington-sponsored initiative did interest d'Arbeloff. A consortium called Sematech was launched in 1987 to explore ways through which the beleaguered U.S. semiconductor industry might legally collaborate to advance its technology and thus its share of the world market. The U.S. Government and the industry each put up $100 million, and Intel founder Bob Noyce, easily the most admired person in the semiconductor industry, signed on as Sematech's President.

Sematech existed to further the aims of the U.S. semiconductor firms, but since these aims depended on the availability of world-class manufacturing gear, a satellite association, called SEMI-Sematech, was formed as a forum through which equipment makers like Teradyne could work cooperatively with the industry. D'Arbeloff, long a strong advocate of dialogue between the chip makers and their equipment suppliers, served as Chairman of SEMI-Sematech for several years — a post that made him an ex officio member of the Sematech Board.

Sales in 1988 grew by more than 20 percent, but the acquisitions were more a hindrance than a help, and $16 million worth of write-offs at CASE, AIDA, and Zehntel drove the bottom line into the red once again. Over the three-year period from 1986 through 1988, the Company had recorded a net loss of more than $36 million. The fact that bookings had turned up provided some consolation. The A500 was a runaway success, its installed base now including AT&T, NEC, Northern Telecom, Siemens, Texas Instruments, and AMD.

The J953 VLSI Test System, now available in several configurations, had captured orders at Delco, Chrysler, AMD, and, most significantly, Motorola. A new memory tester, the J937, defied conventional wisdom by selling reasonably well in the U.S. and Korea. (A U.S. trade magazine had reported the system's introduction with a front cover showing a roulette wheel under the caption, "Teradyne's Big Gamble.")

Despite these victories, Teradyne's U.S. semiconductor customers were still hurting, and the Company depended on its other businesses to support its heavy engineering burden. These businesses were coming through, validating d'Arbeloff's diversification strategy. Teradyne Connection Systems racked up its 18th growth year in its 19-year history, posting record sales of about $75 million.

The Telecommunications Division was contributing more than $60 million to the top line, its profit margins were solid, and it could claim that its 4TEL system was on line testing 31 million telephone lines in the U.S. and Europe. Yet d'Arbeloff was increasingly impatient over what he perceived as the Division's shortcomings: a narrow 4TEL customer base and a failure to break out of the single-product envelope.

Two options suggested themselves. One, favored by d'Arbeloff, was an all-out effort to sell 4TEL to Japan's NTT, possibly via the Cincinnati Bell connection. The other, favored by Ashdown, was entry via acquisition into the world of fiber-optic testing.

In the event, neither road led to Oz. NTT did place a few 4TEL orders, and systems were shipped to Japan for field trials in Tokyo, Nagoya, and Osaka, but that was it. The fiber optics company was not acquired, and instead a modest internal effort was launched. The Company's acquisition plate was full, and in the circumstances d'Arbeloff, no fan of acquisitions to begin with, did not relish the idea of selling another deal to the Board. Feeling stymied and at any rate ready for a new challenge, Division founder Russ Ashdown left the Company in April, 1988. His contributions to Teradyne over two decades were enormous, but his legacy goes well beyond that. Although they will never realize it, tens of millions of telephone users the world over enjoy higher quality service because of the 4TEL systems that were born of his genius.

The shock that reverberated throughout Deerfield moved d'Arbeloff to name as Ashdown's replacement Loren Eaton, Teradyne's erstwhile Personnel Manager. Eaton lacked line-management experience and knew little about the 4TEL business, but d'Arbeloff felt that the first order of business was to stabilize the Division, and that Eaton's people skills would carry the day.

The board test business, despite the distractions of the recent acquisitions, was coming through handsomely. In an effort to exploit the advanced technology incorporated in its large L200-series systems, the Assembly Test Group began offering subassemblies called Digital Test Units (DTUs) to military contractors. An early buyer was General Electric, prime contractor for the U.S. Navy's CASS (Consolidated Automated Support System) program. The catalyst was a Congressional report suggesting that U.S. taxpayers could save millions or even billions of dollars if the armed services purchased commercial test equipment instead of the specialized, very expensive testers then in use.[1] Teradyne's DTUs were just what the Congress ordered, and Teradyne quickly became the leading supplier of off-the-shelf testers to the military, generating sales of well over $100 million over the following decade.

[1]Naturally, the military coined an acronym for this concept: "COTS, for Commercial Off-The-Shelf Testers."

In Walnut Creek, California, meanwhile, the recently acquired Zehntel was determined to prove itself to its new owners. With the ink on the merger papers hardly dry, the unit shipped the first copy of a revolutionary new board tester called the Z1800. This compact, low-cost in-circuit tester was like nothing the board test market had ever seen. The test system controller was a personal computer. (The buyer could use his own.) The programming was via PC spreadsheet. Even the marketing was unorthodox, with direct selling by mail and telephone accounting for many orders. It took Zehntel only one year to sell 150 systems at prices averaging about $75,000, and the acquisition thus gave Teradyne command of both ends of the board test price/performance spectrum, with the Z1800 at the low end and the L290 combinational system virtually owning the high end. It also provided market balance, Zehntel's commercial sales nicely complementing the heavily military sales at the Boston board-test operations.

In 1988, the time came to expand the officer corps again, and two new vice presidents were named: Rick Dyck, now firmly in control of Teradyne's Japan operations, and Kris Nasta, watching over Teradyne's growing 4TEL business in Europe.

In 1988 d'Arbeloff decided to take Teradyne out of the laser trim business. Well over $100 million worth of Teradyne laser trimmers had been sold, including $43 million in 1984 alone. Many of the laser trimmers were sold with memory testers, and the Company effectively exploited its position as the only seller of both tester and trimmer. The chief laser trim competitor, Electro Scientific Industries (ESI) made only the laser trimmers, and none of Teradyne's memory-test competitors offered a laser trimmer.

It was a powerful combination while it lasted, but in the late 1980s the advent of off-line trimming took the bloom off the rose, and there was also the massive shift in memory production from the U.S. to the Far East. While the Company had enjoyed notable success selling trimmers to Mitsubishi, it could not capture a significant part of the Japanese market. It was one thing to stay the course in memory testing, where part of the engineering investment could be shared with the VLSI test group. But the laser-trim business had to stand or fall on its own. D'Arbeloff's verdict: It was time to divest.

Thus, at the end of 1988, Teradyne sold a majority interest in its laser trim business to General Scanning Corporation, the supplier of the beam positioners used in the Teradyne trimmers. Because the Teradyne name had considerable market value, the new operation would for a time be called Teradyne Laser Systems, Inc. (TLSI). Several of Teradyne's engineers and managers left Teradyne to join TLSI.

At the Company's Boston headquarters on Harrison Avenue, an adjacent service station and a building housing a small electrical contractor were purchased, giving Teradyne ownership of the entire block and a chance to build a garage for its employees. The construction possibilities were limited by the fact that the entire block, like so much of downtown Boston, was land reclaimed from Boston Harbor. The fact that part of the land had been used as an automotive service station added environmental factors to the pot. The City, mindful of the fact that Teradyne was its second largest employer (after Gillette), was not eager to throw up so many barriers as to drive Teradyne to the suburbs, but the regulatory climate was still difficult, and Stu Osattin, Teradyne's point man on real estate matters, was soon facing off with the Boston bureaucrats.

Another problem involved the cementitious panels that covered the exterior of the old Hub Mail building. The material proved to be porous, and after four Boston winters the panels had developed cracks, which were bound to get worse. The only option was to replace them all with a new, non-porous skin. That was bad enough; the fact that the old skin contained asbestos turned the whole affair into a nightmare.

The Company persevered, the building was reskinned in 1989, and the garage went up a year later. Direct access from the garage to the Harrison Avenue plant was a real luxury for employees, especially in the winter. Given the fact that the garage and the building were adjacent, this access would seem to be a simple matter. But building is never simple in Boston. Old maps showed a public way between the two structures, and, even though such a road never existed, pipes had been buried underneath its projected path. To this day, those pipes run underneath the lobby structure that connects the Harrison Avenue plant to the garage.

The 300-car garage was greeted warmly by those employees who drove to work. Most of them had been parking in a lot leased from the State, under the Southeast Expressway, a block east of the Harrison Avenue building. This lot was slated to be used as a staging area for Boston's "Big Dig" — the reconstruction of the City's Central Artery — and there seemed to be some urgency behind the garage project. As it turned out, the overflow from the garage continued to use this and other off-site lots for years after the garage was completed.

Employees using the lot under the Expressway had been paying a monthly parking fee of $45 for the privilege. The rationale for the fee: It would be unfair to subsidize employees who drove to work (by granting free parking) but not those who commuted by mass transit. The fee for garage parking was set at $75 a month, in order to help defray the cost of maintaining the structure. Again, the logic was sound, but there was some grousing. Teradyne was, after all, the only high-tech company around that was charging its employees to park, and Teradyne employees working at the Company's plants in New Hampshire, California, and Illinois all parked free. On the other hand, $75 a month was a bargain compared with the parking fees charged by commercial garages in Boston, and anyway, no one was forced to park in the garage. As it happened, there was always a waiting list for any of the 300 spaces that became available.

In 1988, Teradyne captured 14 percent of the world market for semiconductor test equipment. In 1989, the Company added three points to its share. Even so, the math said that five of every six dollars spent for semiconductor test systems went to Teradyne's competitors. A market analysis carried out within the Company suggested that a broad-line test system supplier like Teradyne needed a market share of at least 20 percent to fund its programs and maintain viability.

The problem was clearly not in the mixed-signal area; the A500 doubled Teradyne's share of this market in just two years, and that share was almost double the 20-percent threshold value. In memory test, the major problem remained the Company's inability to crack the Japanese market. While Teradyne could point to sales of over 100 of its new J937 Memory Test Systems, Japan's Advantest was selling *thousands* of systems into its home market.

The Advantest advantage rested on two pillars: Partly owned by Fujitsu, Advantest was a certified member of "Japan, Inc." and could network its way into any company headquarters or board room in its home country. Second,

Advantest was unquestionably a very good company, producing excellent systems widely known for their reliability. The performance advantage often went to Teradyne, but the margin was rarely wide enough to be decisive. Thus, while Japanese suppliers Advantest, Ando (affiliated with NEC), and Minato feasted on the booming Japanese chip industry, Teradyne was forced to hope that the two remaining U.S. producers, IBM and Micron, would stay the course. (Europe was a minor factor in memories.) The only other significant opportunity lay in Korea, where Teradyne had made inroads at GoldStar and was working hard to penetrate Samsung and Hyundai. All other things being equal, the Koreans might favor a U.S. supplier, but it was up to Teradyne to make sure that all other things were at least equal.

In VLSI logic testing, demand for test systems generally waxed and waned with the fortunes of the computer industry, and particularly of the personal-computer makers. Teradyne was now firmly embedded at Motorola and AMD, but the microprocessor giant Intel had slipped out of its grasp, turning to LTX's Trillium Division (the latest reincarnation of the roving band that had earlier staffed VLSI test operations at Fairchild and GenRad) for testers for its 80386. One microprocessor generation later, the Intel wheel would spin full circle, when Schlumberger (ex-Fairchild) would take the business away from LTX.

Board test sales were sluggish in 1989, although the Z1800 continued to thrive in the low end of the market. In the military market, the CASS money machine took a momentary breather. Michael Schmidt returned from Japan to take over as CASS program manager, and Mark Jagiela, an engineer who had joined the Company in 1982, became the new manager of the Japan Division.

Even with the semiconductor industry recovering, Teradyne's sales grew by less than 5 percent in 1989, and net profit was about $10 million, or barely 2 percent of sales. Net income for the five-year period from 1985 through 1989 totaled *minus* $12.3 million, although sales grew from $393 million to $484 million. (These figures are restated to reflect acquisitions.)

Four new vice presidents were appointed in 1989: David Bernstein, managing sales in the Assembly Test Group; John Casey, running semiconductor test sales; Jeff Hotchkiss, managing the EDA effort; and Dan Murphy, Connections Systems sales manager. All were Teradyne veterans save Bernstein, recently hired from a Massachusetts public-sector agency.

The return to profitability, such as it was, may have tempted some at Teradyne to believe that the Company had overcome the adversity of the late 1980s and was back on its growth track again. If anyone did believe this, they were brought face to face with reality in 1990, when the Company posted a net loss of $21 million as sales dropped from $484 million to $459 million. Quarter after quarter, the losses mounted, and the reality of the situation was impossible to ignore: Teradyne had the best and the brightest people in its industry, it had excellent products across a broad line, a good reputation among customers. It was financially strong. It operated in growing, though volatile, businesses.

And it had lost money in four out of the last five years.

Boards of Directors exist, many people believe, primarily for the purpose of appointing and un-appointing chief executive officers. That is an exaggeration, but the thought must have crossed a few minds on the heels of the Company's dreadful performance in the second half of the decade of the 80s. And, while there is no evidence to suggest that d'Arbeloff was under siege, he understandably sought some expression of support from the Board.

His tactic was to issue a challenge to his Directors: Did they believe, as some commentators had written, that the U.S. was going to be forced out of the semiconductor business? If so, he would shrink the Company and look for a viable niche business, a safe haven. Or did they believe that the U.S. would ultimately recover, giving Teradyne the chance to regain the ground it had lost?

Faced with a Hobson's Choice, the Board of course was unanimously in favor of staying the course. But d'Arbeloff knew that the Company's turn-around would require more than Board support, however welcome it was.

19

The Problem Is Us

The Company's lackluster performance over the preceding five years forced d'Arbeloff to a bitter conclusion: There was something terribly wrong about the way the Company was managing its affairs, and strong medicine would be needed to get the ship back on course. One prescription was suggested by Director Ed Artzt, the Procter & Gamble CEO, who offered to send P&G's top quality officer east for a brainstorming session with Teradyne's management. The meeting in Boston was instructive, as it became painfully clear that, no matter how well versed Teradyne's managers were in matters of strategy and finance and general business principles, they were not current in the theory and practice of quality.

Not for lack of trying. Years before, a number of Teradyne's senior managers, including Prestridge, Chamillard, Robbins, and d'Arbeloff, had made the pilgrimage to the then-trendy Crosby "Quality College" in Florida. But the Crosby philosophy leaned heavily on the use of motivational techniques, and d'Arbeloff was more interested in implementation. The Crosby experience raised management's general sensitivity to quality matters, but in d'Arbeloff's opinion it did little more than that.

Now, the Company Nick DeWolf had built by promising customers an unprecedented level of reliability was certainly no laggard with respect to manufacturing quality. Under Chamillard, Production Manager Ron Dias, and others, Teradyne's assembly operations and circuit-board foundries were always driving hard on quality, and the "mean time between failure" of Teradyne systems, while not yet up to the best Japanese standards, was steadily improving.

But d'Arbeloff had bigger fish to fry. Quality, to him, meant not just building more reliable products; it meant building a better company. The crusade he was searching for would necessarily involve everyone, in every corner of the Company.

This search led him, in late 1989, to investigate a new management technique called Total Quality Management. It was also known by the chief benefit it promised: Continuous Improvement. To learn more about TQM, he asked John Petrolini, quality manager for Teradyne's circuit-board foundry, to identify a core body of work — books, seminars, courses, etc. — on the subject. Petrolini gave d'Arbeloff a list of six books and three seminars,[1] and d'Arbeloff promptly gave this list to his senior management team, with the further suggestion that the team meet once a month for a full day to discuss ways in which TQM might be applied at Teradyne.

D'Arbeloff, who had once been a severe critic of quality programs as formulated by Crosby and others, was an instant and enthusiastic convert to TQM-ism, and his two chief lieu-

[1]The books: *The Deming Route to Quality and Productivity; Road Maps and Road Blocks* by William Scherkenback; *What is Total Quality Control?* by Kaori Ishikawa; *Company-Wide Total Quality Control* by Shigero Mizuno; *Quality Without Tears: The Art of Hassle-Free Management* by Philip Crosby; *Juran on Leadership for Quality: An Executive Handbook* by Joseph Juran; and *Out of Crisis* by W. Edwards Deming.

The seminars were sponsored by the American Supplier Institute, Texas Instruments, and W. Edwards Deming.

tenants, Jim Prestridge and Owen Robbins, were just as eager to put the Company on a new course. The only question was one of implementation. A canned TQM program would not do; the employees might see it as another management fad and would lose interest unless they could see quick results.

Teradyne was not the only U.S. company making an agonizing reappraisal of its management practices. Just a few miles south of Boston, Analog Devices founder and CEO Ray Stata, a friend of d'Arbeloff's, was also looking for answers, and his search led him to Professor Shoji Shiba, on the faculty of the University of Tsukuba but at the time a visiting professor at MIT. Shiba suggested the formation of a Boston-area "Center for Quality Management," a forum in which local high-technology companies could travel the TQM road together. He further suggested that the companies appoint a working committee charged with observing TQM first-hand at sites in the U.S. and Japan. The Center was formed, and d'Arbeloff dispatched John Petrolini and Ron Butler, both with extensive manufacturing experience at Teradyne, to make the TQM Grand Tour.

All this went on more or less out of view of the Teradyne work force, and the first public indication that things were about to change occurred at the Company's Sales Meeting in May, 1990. A talk by the Company's CEO was *de rigueur* at such meetings. But this one was different. Instead of reading a prepared talk while standing behind a lectern, d'Arbeloff sat on a stool, dressed in a sweater, and extemporized for 20 minutes, methodically building his case: Teradyne would have to embrace TQM if it was to survive. It was d'Arbeloff at his best, chatting with friends on a subject he felt passionate about, referring to notes from time to time but generally rambling from point to point as new thoughts occurred to him. The talk, which was videotaped, was later shown to many other audiences both inside and outside Teradyne.

Why was this talk so important? Because it established that the coming crusade would be led personally by the Company's CEO. This was not d'Arbeloff showing his troops how to get to the battlefield. It was Patton saying "Follow me!" and charging off to meet the enemy.

During the summer of 1990, the CEOs of the member companies of the Center for Quality Management intensified their efforts to become versed in TQM, first through a one-day seminar with Shiba, then via a six-day course, two days at a time. Watching Shiba in action, d'Arbeloff felt certain that he had found the perfect mentor to help him launch Teradyne's TQM program, and Shoji Shiba was hired as a consultant.

Eventually, a Teradynized TQM program took shape in d'Arbeloff's mind. At the outset, it would focus the Company's entire energies on a single element out of the many in the TQM toolset: the PDCA or Seven-Step program, TQM's formulation of the scientific method. (PDCA stands for "Plan-Do-Check-Act.") PDCA virtually guaranteed that employees would see results soon and would thus be motivated to continue the program.

The new culture would infiltrate the Company from the top down, with d'Arbeloff and five or six senior officers downloading the material to the 30 or so managers who reported to them, these managers training the next layer, and so on until everyone was aboard. Over-all program direction would be entrusted to one of the Company's key middle managers for a one- or two-year term, and while these managers rotated in and out of this position a permanent TQM professional would provide continuity. Mike Bradley, who had made his mark as

A500 Marketing Manager, was the first TQM Manager, while John Petrolini settled in for a long run as Teradyne's Director of TQM Technology.

By the fall of 1990 Teradyne's key managers had all read the prescribed books and had attended the prescribed seminars. There was a dash of skepticism, but it was quickly swept aside by d'Arbeloff's obvious intention to push TQM with all the fervor he could command. Teradyne was going to change radically, and he was setting a course that would keep it changing until it got it right. Any Teradyne managers who resisted did so at their own peril.

PART FIVE

The Nineties. Renaissance.

20

• • •

D'Arbeloff Launches His TQM Campaign

The history of Teradyne to this point consists of a series of challenges: the challenge of starting a new company and inventing a new market; the challenge of transforming Teradyne from entrepreneurial venture to successful mid-sized company; the challenge of recovering from the loss of a group of key engineers and managers; the challenge of meeting the new threat from Japan; and now, the challenge of overhauling the Company's management processes, from top to bottom, in accordance with the principles of Total Quality Management. The first thirty years had produced a half-billion-dollar enterprise that sat atop the industry it created. Now its founder was saying that was not good enough, and the facts were all on his side.

The decision to bring TQM to Teradyne stands as a watershed event in the Company's history. The facts speak for themselves: Teradyne lost money in four of the five years from 1986 through 1990 inclusive. The TQM program was launched in 1991. From that point through the end of the century, the Company was profitable in every quarter of every year. The success can't be credited entirely to TQM — the 90s were generally a time of plenty for the semiconductor and semiconductor equipment industries — but only a fool would believe that the abrupt change in Teradyne's fortunes in 1991 and its embrace of TQM at the same time was a happy coincidence.

Of course, in 1990, as the Company prepared to close the books on a dreadful year, none of this could have been foreseen, not even by d'Arbeloff as he played his role as TQM messiah. Record orders in the fourth quarter encouraged the belief that 1991 would be a decent year, but the dismay over the Company's poor performance in the late 80s hung over Teradyne's managers like a heavy cloud, and d'Arbeloff saw in this dismay the catalyst he needed to drive home the importance of fundamental change.

Thus in February he summoned 30 of Teradyne's top managers to Boston for two days of soul-searching. Why was the Company doing so poorly? Was it serious about profits and growth, or was it looking for excuses to rationalize its wretched performance? Did it intend to take the steps, however painful, necessary to return promptly to profitability? On the first day, the options were laid out and placed on a ballot. On the second day, the managers voted: No more alibis; the Company was going to become profitable, period. Out of this meeting came some harsh measures: two-week plant shutdowns, deferral of the usual mid-year salary increases, and cuts in capital spending and the travel budget.

Of course, renaissance does not come just because managers will it so. And good intentions do not sweep away the hard tradeoffs between short- and long-term goals. But the meeting was valuable as a collective acknowledgment of the need for change. D'Arbeloff could not have asked for a better launching pad for his TQM campaign, in which, at the time of the meeting, about 300 of Teradyne's employees had already enlisted. Soon dozens of

"QITs" (Quality Improvement Teams) were springing up to attack problems as diverse as the high down-time of Harrison Avenue's elevators and the purchase of stationery supplies.

As the teams completed their projects (or, in the TQM vernacular, as they "spun their PDCA wheels") and saw tangible results, enthusiasm grew. TQM had arrived at Teradyne with a bang, and it had taken root. The circuit-board foundry in Boston led the way, but Nashua, with Windy Hunter as TQM Manager, was not far behind, and the other divisions fell in step with varying degrees of spirit. It was perhaps too much to expect the EDA groups[1] to join in, and in any event Case and AIDA, the two CAE companies acquired in 1987, had other things (like survival) on their minds.

In 1990 Teradyne Director Dan Gregory left to accept the position of Secretary of Economic Affairs in Massachusetts Governor Weld's administration. Gregory had joined the Board in 1977 succeeding William Elfers as Greylock's representative, and his loss was significant — but, fortunately, short-lived. A year later he left State government and returned to the Teradyne Board.

As if scripted, recovery came on the heels of the TQM launch. Sales in 1991 rose about 11 percent, to $509 million, and net income was more than $18 million — a $40 million positive swing versus the previous year's net. Most of the sales gains came from the semiconductor test and board test operations, while Connection Systems experienced a rare sales decline (though the Division's bookings and its operating income grew). A new VLSI Test System, the J971, followed its predecessor the J953 into Motorola, to test that Company's leading-edge 68040 microprocessors.

[1]The term "CAE" — Computer-Aided Engineering — had been replaced in softwarespeak by "EDA", for Electronic Design Automation.

A new memory tester, the J990, combined ECL and CMOS circuitry (Teradyne coined "E/MOS" to describe it), along with more timing generators, some powerful new software, and an architecture promoted under the term "tester per site." IBM, one of only two remaining volume memory producers in the U.S., embraced the new system — as it had the earlier J937. Long gone were the days when Teradyne found it impossible to gain an audience at IBM; the computer giant was now regularly among Teradyne's largest customers.

In circuit-board test, the Navy's CASS program was spending tens of millions of dollars on Teradyne subsystems, and the Z1800 passed the 800-unit mark. In an effort to spur Far East sales, Teradyne signed a partnership agreement with a Japanese maker of circuit-board handlers, Okano. It was a well-intentioned attempt to find out whether Teradyne might ride into the Japanese board-test market inside a Trojan horse, but it would amount to little.

The EDA business resisted the best efforts of Lassiter and Hotchkiss to make Teradyne a player in that market. Sales at the Telecommunications Division were stalled, as the 4TEL group looked for new worlds to conquer. An obvious target was Deutsche Telekom, and Teradyne opened its German campaign by signing joint agreements with two local equipment suppliers, Vierling and KTS.

Over-all, despite the mixed results, there was more than enough good news at Teradyne to convince most employees that the TQM program was working, and the Quality Improvement Teams kept spinning their wheels enthusiastically.

Any doubts that Teradyne had turned the corner vanished in 1992. The Company's sales grew to a record $530 million, with net income up 24 percent, to $22.5 million. Partly because of a severe slowdown in Japan, the world-wide market for semiconductor test equipment was roughly flat over the two-year period of 1991-1992. Over the same period, however, Teradyne's semiconductor test sales *increased* by 40 percent. The market-share gains were breathtaking: In VLSI test, on the strength of the J971, the Company gained six full points of market share. Three points were picked up in memory test, and in the linear/mixed-signal market the A500-series drove Teradyne's market share above 40 percent for the first time.

Connection Systems, whose Peter Kelley had built a successful operation in Tokyo, now turned its sights on Europe, building a plant in Dublin and moving Kelley to Bracknell as head of International (Europe and Asia) sales. The Division was back on its accustomed growth track, posting record sales in 1992 and introducing a new high-density card connector trade-named HDM. The HDM connector and its subsequent derivatives would fuel the Division's growth for many years to come. For the year, TCS managed to squeeze a 26-percent increase in operating profit out of a 6.4-percent increase in sales — a feat whose significance was not lost on those sizing up George Chamillard's management skills.

Much of the Company's success could be credited to TQM. Manufacturing and test costs for new systems like the new, top-of-the-line A580 were driven down dramatically, and the "Voice of the Customer" and "Market-In" programs, straight out of the TQM song book, were convincing major customers that Teradyne had changed its stripes, and that, if there ever had been any Teradyne arrogance, it was long gone. To drive the point home, Teradyne even invited customers to participate in its QITs. Most important of all, Teradyne began integrating the resources assigned to each major customer, so that the sales, service, and applications specialists assigned to, say, IBM all became elements of the IBM "customer team," whose sole objective was to ensure that customer expectations were always met and often exceeded.

Another sign of the new Teradyne was a burst of technology-sharing across the semiconductor test divisions. The VLSI and memory systems, both produced at the Agoura Hills plant, used the same architecture, the same (E/MOS) components, the same operating system. The Industrial/Consumer Division in Boston was able to import digital test technology from Agoura Hills and to export linear test technology to the Japan Division, whose new A563 LCD Test System was introduced in 1992.

The Industrial/Consumer Division also imported technology from outside the Company where it made sense. Several years earlier Teradyne had purchased the assets of a small competitor, Attain Corporation, and in 1992 ICD introduced a new test head designed around the Attain technology. Called PATH (for Production Analog Test Head), it was designed to optimize linear-IC testers for use with handlers and probers.

Sales at the Telecommunications Division were down about 5 percent in 1992, but hope grew out of the signing of sales agreements with Philips in Europe, plus two companies in South America. And the courting of Deutsche Telekom continued. The Philips pact quickly led to 4TEL sales in the Czech Republic and Slovakia, and by year-end the 4TEL line count reached 48 million, in 13 countries. A new fiber optics test system, hopefully called "4SIGHT," was scheduled for completion during the year, but as it began to appear that 4SIGHT might be too little, too late, d'Arbeloff grew antsy again. As he saw it, the Division had lost momentum.

In board test, the Z1800 had moved Teradyne into the front ranks of the commercial market, with 800 units shipped by year-end. The line was extended by the addition of a Vector Processor, designed to test boards with ASICs and microprocessors, and a system that came equipped with the Okano mechanical fixture. But total board test sales sagged in 1992. Part of the problem was weaker demand for the high-end combinational test systems. The hard fact was that it took too many Z1800s, at about $75,000 apiece, to make up for the loss of a dozen L300s, each of which could bring $2 million or more.

Prospects were still bright in the military board-test market. By the end of 1992, Teradyne had shipped a total of 92 CASS systems, out of a projected total of more than 700, and the Company was able to propagate CASS technology to other systems, subsystems, and modules, so that it in effect owned the franchise for digital test technology employed in military systems. To solidify this position, Teradyne in 1992 joined with two other companies involved in the CASS business — prime contractor General Electric and subsystem supplier Hewlett-Packard — to form the Automated Test International Limited Partnership, a new enterprise in which the three companies would hold equity.

Despite the success of the Z1800 line and of Teradyne's thrusts into the military market, the board test business was a riddle Teradyne had yet to solve. Part of the problem was a total market that rarely grew by as much as 10 percent a year, and part was Teradyne's inability to gain a meaningful share of the commercial market. That market arguably offered a reasonable return for the market leader and perhaps for the number two player, but in 1992 Teradyne held neither of these positions. As for the military market, where Teradyne reigned supreme, the Cold War was over, and politicians eyeing the "peace dividend" would almost certainly start slashing the military budget soon.

Board test was not the only frustration for Joe Lassiter. The two EDA companies acquired in 1987, Case Technologies and AIDA, had not lived up to expectations, and their key personnel began to migrate to their next ventures. So Teradyne liquidated these holdings as best it could and circled its wagons around the LASAR group in Boston. This group introduced a new software package (VICTORY) designed to facilitate the use of a board-test technique called Boundary Scan, and it also produced a software product called BoardWatch, which electronics manufacturers could use to monitor the quality of their circuit boards. As usual, the technology was first-rate — Teradyne always had a stable of talented software engineers — but converting that technology into a profitable, growing business was another matter.

Rick Dyck's attempts to break into the Japanese market for logic testers were beginning to bear fruit. Just as the Japanese ATE makers had benefited from their country's dominance of the memory market, Teradyne benefited from the U.S. dominance of the microprocessor market. In the course of serving U.S. microprocessor giants Intel, Motorola, and IBM, Teradyne was developing a technical edge that Japanese chip makers could not afford to ignore. The breakthrough came when Japanese chip makers Toshiba and NEC purchased licenses to use microprocessor technology developed by MIPS, a California company and a user of Teradyne's J971 VLSI tester. The test programs had been written, the applications work had been done, and for the Japanese the selection of a VLSI tester was a no-brainer, and the J971 rode into Japan on the wings of the MIPS chip. And one Japanese specialty market belonged entirely to Teradyne: Sony had long relied on Teradyne for its image-sensor testers, and now this market was exploding, thanks to growing demand for video cameras.

College recruiting, always a high priority in the U.S., was infinitely more difficult in Japan. Rick Dyck, fluent in the language and at home in the culture, was the ideal recruiter, but the graduates of Tokyo's best universities were tradition-bound to work for Japanese companies, and the few renegades were more likely to head for U.S. giants like IBM and Motorola than to smaller companies like Teradyne.

Dyck huddled with Japan Division Manager Mark Jagiela to ponder the problem. Both had traveled often to the Japanese island of Kyushu, home to about a third of Japan's semiconductor production. Might Kyushu, 500 miles away from the competitive bustle of Tokyo, provide the perfect solution? Why not give it a try? Specifically, why not stake a claim in Kumamoto, a city big enough (650,000) to provide the necessary services and home to a number of first-rate engineering schools? The graduates of these schools tended to move to Tokyo or Osaka to find work as designers. A Teradyne site in Kumamoto, Dyck and Jagiela guessed, might persuade many of these engineers to stay in Kyushu. Best of all, Teradyne would be the first U.S. company to locate in Kumamoto, and it could expect to be welcomed with open arms.

To test the thesis, Teradyne leased space in Kumamoto, and employees of the Japan Division began to trickle south from Tokyo.

In 1992, d'Arbeloff turned 65, the age at which many people retire. "Retirement" was not in his vocabulary, but facts had to be faced: His two top officers, Prestridge and Robbins, were also both in their sixties, and the Company would soon be hit with a wave of retirements. (The process indeed had already begun, with the retirement of Dennis O'Connell.) Always thinking ahead, d'Arbeloff began moving pieces around the chessboard. In 1992,

Prestridge and Robbins were named Executive Vice Presidents, and both were appointed Directors of the Company. Mike Bradley, having served his tour as TQM Manager, was named Vice President in charge of North American semiconductor test sales, reporting to John Casey, and Tom Pursch moved from Connection Systems into the TQM seat Bradley vacated. George d'Arbeloff became head of the Semiconductor Test Group, with Sulman in Boston and Rogas in California reporting to him.

The most telling of d'Arbeloff's chess moves occurred early in 1994, when George Chamillard was brought from TCS to Boston and given the title Executive Vice President, joining Prestridge and Robbins on a newly formed Executive Committee. The significance of Chamillard's promotion was clear enough to most people, but, in case there was anyone on the planet who didn't get it, d'Arbeloff wrote, in his 1993 Annual Report letter (published in April, 1994), "With George's appointment we... have preemptively addressed the issue of management continuity."

The fall-out came quickly. Joe Lassiter, a plausible candidate for the top job, left the Company. Jeff Hotchkiss succeeded him as manager of the Assembly/EDA Test Group.

More moves were in the air at the Telecommunications Division, in Deerfield, Illinois. Loren Eaton had worked hard to smooth the troubled waters in the wake of Russ Ashdown's departure, but the Division stalled noticeably in 1991 and 1992. John Casey, the Company's semiconductor test sales manager, was sent to Deerfield as Vice President for Sales and Marketing, taking over as Division Manager later in the same year.

Casey by this time was well recognized as one of the Company's ablest managers. Tall and handsome (a Christopher Reeve double), he was also extremely conscientious and well liked. As world-wide sales manager, he had developed a encyclopedic knowledge of the semiconductor test territory, and customers had come to know that whatever Casey promised, Casey delivered.

The departure of such a major talent from the semiconductor test scene, in order to run a relatively small division, raised a few eyebrows. But the Telecommunications Division needed help, and Casey was a good soldier. Second, Mike Bradley, his back-up in semiconductor test sales, was a highly capable replacement. Third, the loss would not necessarily be permanent; Casey could later be recalled to the semiconductor test wars (and in fact would be).

The pieces were now falling into place for an orderly management transition, which would occur over the next few years. Chamillard would become President and Chief Operating Officer in January, 1996 and Chief Executive Officer in 1997. Four short years after d'Arbeloff tapped Chamillard as his successor, the process of succession was complete. Few companies worked it all out as smoothly, and a great many companies pulled it off badly. The business press is full of stories about CEOs attempting to hold onto power after their time had passed and about the convulsions caused by their intransigence.

Casey's arrival at the Telecommunications Division was followed by the largest single order in the Company's history: a \$63-million contract to equip the entire German telephone network (Deutsche Telekom) with 4TEL systems. Obviously, much of the groundwork for this order had been laid by Loren Eaton in the course of many trips to Germany over the preceding two years, but the order was closed on Casey's watch, and it was momentous, since it stamped Teradyne as the hands-down line-test leader on the Continent and would certainly lead to substantial follow-on business. Some of this came quickly, as the

Czech Republic ordered a second 4TEL system, and as Hungary's telephone company began looking seriously at 4TEL.

The Deutsche Telekom order was far from the only good news in 1993. The Company once again posted record sales ($555 million) and net earnings ($35 million). For every dollar of added sales, the Company brought about 50 cents down to the bottom line. Over the three-year period from 1991 through 1993 (essentially the period since the TQM program began) net profit had increased by $56 million on a $96-million sales gain. The good times were back, big-time. With cash building fast, the Company spent $10.8 million to redeem the convertible debentures it had inherited with the acquisition of Zehntel and began to purchase one million shares of its stock on the open market.

In semiconductor test, both the east-coast and west-coast operations were growing fast. The A580 was a blockbuster success, and many of these systems carried price tags of over $2 million. The third floor of the Harrison Avenue building was outfitted as a parade ground for the large A500 systems, where visitors could see each system, wearing the name of its future owner, being assembled and tested. At this point the A500 line, with barely five years under its belt, was one of the most successful products in the history of the ATE industry, with more than 300 shipped at an average selling price of more than a million dollars.

The J900 line of memory testers series included, predictably, three systems, for engineering, wafer probing, and final testing. The U.S. market for such systems had essentially narrowed down to IBM and Micron, but at both accounts Teradyne was the dominant supplier. Teradyne systems were also on many of the test floors of Korea, but the all-important Japanese market still belonged to the Japanese ATE companies, led by Advantest.

In VLSI logic test, the J971 continued its winning ways among microprocessor makers (with the notable exception of Intel), and another new VLSI test system, the J921, targeted the growing market for systems to test RISC processors and ASICs. The J921 sported the latest variation of the tester-per-pin architecture pioneered years before by Megatest, this one called "palette-per-pin." The system also used gallium arsenide circuits to drive speed to 150 MHz.

Connection Systems brought in the largest order in its history, for $12.5 million worth of backplanes and connectors, for Motorola's European cell phone operations. The order was significant on several counts: It drove the Division's commercial business to 60 percent of the total, proving that the military-to-commercial transition had been pulled off brilliantly. It increased non-U.S. sales to more than 15 percent, and it validated the decision to build a manufacturing plant in Dublin — the planned site for most of TCS's Motorola production.

TCS by 1993 had become one of the world's leading (if least recognized) connector companies, with annual sales approaching $100 million. It had achieved this status by carefully remaining in its tightly prescribed niche, the supply of mostly large, multilayer backplanes ("mother boards") equipped with the high-density, high-performance connectors on which the performance of cutting-edge electronic equipment heavily depended. A premium price was charged for a premium product, and any thoughts of downscaling to expand the served market were quickly dismissed. But now the market itself was moving upscale to meet Teradyne, and pressure grew to respond to the much larger market that came knocking at the door in Nashua.

The response took a variety of forms, as the Division began to move tentatively toward the broader EMS (electronic manufacturing services) market that was emerging as equipment makers decided to subcontract the manufacture of their products — to buy rather than build. The challenge, as always, was to become a player in the high-end market while steering clear of the highest-volume, commoditized sector when margins were paper-thin. And so a new TCS business strategy took shape: The Division would stay firmly anchored at the high end of the technology spectrum but would adopt more flexible marketing policies with respect to second-sourcing and to supplying backplanes equipped with non-Teradyne connectors. Any incursions into the EMS market would be carefully orchestrated to emphasize the connector and backplane technologies that differentiated TCS from the pack.

In the wake of Chamillard's departure, TCS welcomed its new manager, Ron Dias. Dias had spent most of his long Teradyne career in manufacturing, with a five-year detour into Human Resources. Now he was back running a Division where production was king, a fish returned to water.

In board test, the annual CASS installment came in on schedule, but reductions in military spending took their toll on L300 sales. The Z1800 line was now one of the Company's all-time best sellers, with over 1000 units shipped, and the system was making inroads into the Asian market, where its low cost and ease of use were decisive factors in its favor. In fact, it appeared likely that the Z1800 had outsold all other in-circuit testers in China. The total was under 100 systems, but the potential was obviously vast.

The Board of Directors traded one Harvard dean for another in 1993, as Albert Carnesale, Dean of Harvard's John F. Kennedy School of Government, replaced John McArthur, who resigned owing to the press of other business.

The recovery that began in 1991 and continued in 1992 and 1993 accelerated in 1994, as sales increased by 22 percent, to a record $677 million. Net income more than doubled, to $71 million, the margin on sales topping 10 percent for the first time since 1984 and only the second time since 1967. Bookings for the year exceeded shipments by more than $100 million, and it was clear that new capacity would be needed. A 120,000 square-foot building in Agoura Hills, mothballed since the previous cycle, was a godsend for the digital IC test divisions, whose VLSI and memory testers continued to gain market share. In Boston, orders for the A500 line soared an incredible 65 percent above the previous year, as Teradyne left its mixed-signal competition in the dust.

True to form, the Company plowed the revenues back into ever-higher engineering budgets, and new A500 options proliferated at Harrison Avenue. The Japan Division, readying a new consumer-device tester to go with its LCD and CCD lines, broke ground in Kumamoto for a new plant to replace the leased quarters it had occupied in that area for several years.

Of all the Teradyne divisions, the greatest sales gain occurred at Connection Systems, up 49 percent to $132 million. The HD+ and UHD connectors were on fire, and a new connector called HDM (High Density Metric) was capturing design-ins by the dozens, planting the seeds for further growth. Sales doubled in Europe, largely on the strength of sales to Motorola's cellular operations.

A major factor in the explosive sales growth at TCS was the decision to offer its unique connectors as connectors rather than simply as components of backplane assemblies. The available market for these proprietary connectors was now wide open, and Teradyne ran with full throttle.

In the board test sector, the Company split the Assembly Test Group into two divisions: Assembly Test Boston (ATB), serving the military market, and Assembly Test Walnut Creek (ATWC), addressing the commercial market. The former would be managed by Jeff Hotchkiss, the latter by Ron Butler. Each of the newly liberated divisions seemed revitalized by the move. At ATB, CASS business picked up steam when Martin-Marietta (taking over from GE as prime contractor) approved a new, VXI-based version of the digital test units, extending the program's term through 2005. At ATWC the Z1800 installed base now included 1300 systems in 500 plants in 36 countries. The Walnut Creek operation gave its star product two new options: the Safecracker Pattern Generator and the MultiScan Vectorless Programming Toolset. The strategy was clear: Fill the world with Z1800s, then make a living off their care and feeding.

The Telecommunications Division, now under John Casey, shipped its first 4SIGHT system (the product was, for all practical purposes, stillborn) and began shipping the huge complement of 4TEL systems covered by the Deutsche Telekom contract. The German company also awarded Teradyne a new $9 million contract to develop software linking 4TEL with the Company's management systems.

The good news resonated on Wall Street, and the market value of Teradyne stock flirted with the billion-dollar mark throughout the year. Teradyne old-timers, who for years had waited patiently for their ships to come in, were finally accumulating real wealth. For those who were approaching retirement, the timing could hardly have been better.

At the top of the list of accumulators, of course, was Alex d'Arbeloff, who had held most of his Teradyne shares through thick and thin. In addition, his prowess in turning Teradyne around and his far-reaching network of contacts in the venture capital community (he was a frequent lecturer on entrepreneurship) resulted in his appointment to the Boards of a number of high-tech companies, including Stratus Computer, Lotus Development, and Thermo Electron. With these directorships usually went the chance to buy founders' stock, and the d'Arbeloff fortune mounted.

In 1994, Teradyne was hitting on all cylinders in semiconductor test. In mixed-signal test — the fastest growing sector — Teradyne's sales exceeded those of its next three competitors combined. In memory test, the Company's short suit in semiconductor test, nine of the world's top 20 memory producers were using J990-series systems. In VSLI logic test, the best-selling J971 was now available with 200-MHz performance, and a late-year surge of orders promised an even better year to come. Over-all, the Company was once again the preeminent maker of semiconductor test systems in the world — *despite* the fact that it could claim neither Intel nor "Japan, Inc." The Intel franchise was held by Schlumberger, and in Japan, where Teradyne was beginning to establish a beachhead in VLSI, the volume of memory testers sold by Advantest, Ando, Minato, kept Teradyne's share of that market irritatingly low.

The U.S. competition, as usual, consisted mainly of Schlumberger and LTX (now trailing badly in the mixed-signal field), plus two relatively recent entrants. One was the instrument powerhouse Hewlett-Packard, which had made feeble sorties into semiconductor test over the years but which now seemed to be readying a real threat, winning a key order for testers at Motorola. Motorola was still Teradyne territory, but no one at Teradyne was taking a competitor like Hewlett-Packard lightly.

Another competitor starting to attract attention was Credence Systems, which had been assembled from odds and ends of the ATE industry, including the remnants of Tektronix's semiconductor test operation and two small ATE companies, Axiom and ASIX. No one had ever built a successful ATE company through acquisitions, but Credence skillfully demonstrated that it could be done. Its president, David Meese, had worked for Teradyne years before, and, when he was tragically killed in a plane crash, Woody Spedden, another ex-Teradyner, replaced him.

Another U.S.-based competitor was Megatest Corporation, with 1994 sales of about $100 million, mostly in the VLSI test segment.

In 1994 John McCabe, the Company's Controller for almost 20 years, turned his spreadsheets over to Don Hamman, the Company's Director of Corporate Accounting, and became Teradyne's first Vice President for Information Systems. Frank Carroccia, Operations head at TCS, was also made a Vice President. In the same year Henry Watts, Teradyne's longest-serving Director save only d'Arbeloff, retired at 89. In recognition of his years of dedicated service Watts was named Teradyne's first and only Honorary Director.

The Teradyne volcano that had been rumbling in the early 90s erupted in 1995, as sales shot through the billion-dollar mark. The numbers, excluding an acquisition that came late in the year, showed a 56-percent increase in shipments and a 122-percent gain in net income.

Semiconductor test sales jumped by an astonishing 75 percent. The market itself was very robust, and Teradyne's competition joined in the party, but Teradyne was gaining share across the board. In VLSI logic test Teradyne gained about four points of market share. The Industrial/Consumer Division, running fast with the A500 line, topped a good 1994 by 56 percent, and the Japan Division, now berthed in its new Kumamoto home, saw its sales soar 80 percent. But the prize in semiconductor test was won by the memory test division, where shipments rose by 120 percent while orders grew even more. The J990-series was making its mark in the U.S., Europe, Korea, and Taiwan, proving, for that year at least, that a viable memory test business could be built without a strong Japanese presence.

Board test sales grew by 30 percent, outpacing the market by 10 percentage points. The Z1800, now with a full complement of options and a 5000-pin version for large boards, continued to thrive in the commercial sector, and in the military, the CASS contract was the cherry atop a sundae that included a long list of weapons systems: the Comanche helicopter, the F-14D, F-15F, F-16 and F-18 fighter programs, and the Patriot, Harpoon, Hellfire, Exocet, and Javelin missiles. But the headline-maker for the year was none of the above; it was a $56 million contract from Northrop-Grumman to equip a test depot for the B-2 Stealth bomber.

Connection Systems sales of $155 million were "only" 17 percent above the previous year, but much stronger growth in Europe and Japan now put overseas sales at about one third of the total — a smaller percentage than the 40-plus percent routinely delivered in semiconductor test, but quite respectable nonetheless. The HDM line was now the computer industry's favorite high-density connector, and, with demand building, TCS looked for expansion space. A 78,000-square-foot-plant near the TCS headquarters, formerly occupied by Harris Corporation, came on the market, and TCS swooped. It also expanded its printed circuit facility (occupying the site of the original TCS Nashua plant) by 15,000 square feet and leased a 15,000 square-foot plant in the Dallas area.

At Deerfield, the Telecommunications Division was rightly feeling upbeat about its German contract, and work accelerated on a new, 4TEL II line test system equipped with expert systems technology that enabled it to use its cumulative experience to produce faster and smarter test routines. Since the 4TEL installed base (90 million lines, including the German order) could be field-upgraded to 4TEL II status, a huge new market beckoned.

In Japan, the Kumamoto experiment was judged a major success. Teradyne could indeed hire the local engineering talent it needed. So, in April, 1995, Teradyne dedicated its new plant on Kyushu, by this time widely known as "Silicon Island." As expected, the officials of Kumamoto were overjoyed to have been "discovered" by a U.S. manufacturer, and the Governor of the Prefecture, a member of one of Japan's leading families, hosted an elaborate luncheon at his estate in honor of the occasion. Teradyne was to do well in Japan, but Governor Hosokawa was to do even better, soon becoming Japan's Prime Minister. His successor in Kumamoto, Governor Joji Fukushima, was equally supportive of the Teradyne presence, traveling with his wife on a "state visit" to the Company's Boston headquarters.

21

A Time for Planting

When asked to name the single most important talent he brought to Teradyne, Alex d'Arbeloff is likely to refer to his penchant for pushing the Company in new directions. On the surface, this doesn't seem to stand up, since Teradyne in 1995 received most of its revenue from products that could be viewed as logical extensions of the semiconductor testers of the 1960s. Moreover, the Company's one famously successful business outside ATE — connection systems — was pursued despite d'Arbeloff's initial indifference, though he claims credit for pushing TCS into the commercial sector that now accounts for most of its business.

And yet, there is no question that throughout Teradyne's history d'Arbeloff exhibited an impatience with the status quo, both by goading his managers to explore new techniques (like TQM) and by pushing the Company's product-line envelope. Even when to all outward appearances Teradyne was an eminently successful company, d'Arbeloff was always looking beyond the smooth water and finding icebergs. He was forever pushing his people into new ventures, and when, inevitably, some of them failed, no one was scolded for making the effort.

The real victims of d'Arbeloff's wrath were those who insisted on playing it safe. A long list of Teradyne-sponsored ventures, some successful, some not, attest to d'Arbeloff's incessant stirring of the pot.

In 1995, d'Arbeloff was at his venturesome best, as he created one new venture to take Teradyne's semiconductor test business in a wholly new direction, a second to take the Company into the mechanical handling business, and a third to launch it into the software test area. As if that weren't enough, he would also preside over the acquisition of one of Teradyne's long-time ATE competitors.

Software Test

It is a truism in the high-tech industry that companies spend as much money developing software as they spend developing hardware. As d'Arbeloff counted noses at Teradyne, he found that the Company indeed employed as many software engineers as it did hardware engineers. If hardware testing (ATE) could become a multi-billion-dollar industry, why not software testing? Most software testing was currently being performed by in-house talent at the major software developers — just as the early ATE was developed by the chip makers' in-house groups. A few companies dabbled in the area, but none seemed clearly on a path to become what Teradyne became in hardware testing. The more he thought about it, the more irresistible the idea became.

In 1994, Digital Equipment was in the throes of dismemberment, and among the expendables was a small group of engineers who had been working on a project to develop a software test system. The DEC team, under manager Ron Kita, was attempting to refine technology Kita had earlier developed at AT&T. Hearing about the group, d'Arbeloff led a small group of Teradyne engineers on a scouting expedition to DEC. Larry Apfelbaum, then a business manager in Teradyne's board test group, saw great promise in the technology and suggested that it might be useful as an adjunct to Teradyne's military board test business. Board test manager Jeff Hotchkiss was cool to the idea, and Apfelbaum then drafted a business plan, based on the technology's commercial potential. D'Arbeloff was sold, and the entire seven-person DEC development team was hired and installed in Nashua as Teradyne's Systems Software & Test (SST) Group, under Apfelbaum.

Anxious to acquire critical mass, d'Arbeloff immediately began looking for other software test companies. His search soon led to Hammer Technologies, a Boston-area company that specialized in the testing of computerized telephone systems. Hammer had been founded in 1991 as Transaction Environments and later adopted the name of its core product. It was a successful company at the time of purchase, with 50 customers, including Fidelity Investments, Walgreen, and all seven of the Bell operating telephone companies. As soon as Hammer was acquired, d'Arbeloff started looking for other targets.

The Integra Story

Teradyne's 1995 success in semiconductor test, incredible on its face, was not quite all that it seemed to be. The total world-wide market for semiconductor test equipment was up 65 percent in the year — less, to be sure, than Teradyne's growth, but not that much less. Credence was growing even faster than Teradyne, not because it was stealing share (the two rarely competed head-to-head) but because it was exploiting a sector so far ignored by Teradyne. On the strength of technology gained through acquisition, Credence was offering low-priced test systems based on the use of compensated CMOS technology.[1] These systems, flying low under Teradyne's radar screen, picked off a growing share of business for which Teradyne's large, relatively expensive systems could not compete. This did not seem to bother Teradyne's semiconductor test organization, which was, after all, selling its million-dollar systems as fast as it could produce them. But it bothered d'Arbeloff greatly. And when Motorola, a long-time, loyal Teradyne customer, revealed that it had narrowed its bidders' list for a new microcontroller tester to four companies — *not* including Teradyne — the message was clear: Teradyne desperately need a new tester, a tester that exploited new technologies in order to deliver a new level of performance per dollar.

D'Arbeloff took his case to the VLSI Test Division in Agoura Hills and the Industrial/Consumer Division in Boston. The answer was the same in both places: The necessary resources could not be spared without jeopardizing the systems that were Teradyne's bread and butter. Teradyne, d'Arbeloff found, had become a prisoner of its own success. It was facing what Harvard Professor Clay Christensen[2,3] would call "The Innovator's Dilemma," the inability of legacy-bound companies to respond to what Christensen called disruptive market shifts.

But d'Arbeloff would not be put off. If the existing organization could not or would not do the job, d'Arbeloff would create a new unit, wholly outside the existing framework, to produce the new tester. To run it, he chose Marc Levine, an

[1]CMOS (complementary MOS) was cheaper and less power-hungry than the bipolar parts used by Teradyne and others in its testers, but suffered from heat-induced inaccuracy. The challenge was to design circuits that would compensate for the effects of rising temperature.

[2]Christensen, Clayton M., "The Innovator's Dilemma. When New Technologies Cause Great Firms to Fail." Harvard Business School Press, 1997.

[3]"How Teradyne Solved the Innovator's Dilemma," *Fortune*, January 10, 2000.

MIT graduate (with a Master's from Stanford) who had joined Teradyne in 1980 and had been ICD's software engineering manager until 1994, when he was named Company-wide TQM Manager. In this role he reported directly to d'Arbeloff and could thus be easily recruited to run the new venture.

Would Levine take on the job of leading the new, entrepreneurial venture d'Arbeloff wanted? Yes he would, if he could have Hap Walker, a gifted engineer and system architect who had come to Teradyne in 1977 from Intel and who had since been one of Teradyne's most prolific designers. It was a deal, and Levine and Walker began crafting their new enterprise, which they named Aurora.

Aurora would begin its life in Boston, where Levine and Walker assembled a small team, some, like software engineer Dan Proskauer, from Teradyne, others, like analog engineer Ron Sartschev, hired from outside the Company. In September the group moved to leased quarters in Bedford, Massachusetts.

The Aurora project engaged d'Arbeloff's enthusiasm fully. He was in effect creating a new company, a new Teradyne, out of whole cloth. It would have its own board of directors, consisting of Levine, Walker, d'Arbeloff, Jim Prestridge, and Ed Rogas. Aurora's manager, Marc Levine, would report directly to this Board, bypassing the Company's existing semiconductor test organization.

The marching orders, broadly speaking, were two: First, the team would design its new system around compensated CMOS. Second, the system would use the Windows NT operating system. Hap Walker approached the first challenge with confidence, the second with misgivings. Like most test-system architects, Walker was a true believer in the UNIX operating system, and he wondered whether Windows was up to the job of running a high-performance test system. (The Z1800 board tester had proved the PC a workable test system controller, but IC testing was in a different league altogether.)

D'Arbeloff, however, was adamant. He was sitting on the Boards of two Boston-area companies[1] that were exploiting Windows capabilities to the hilt, tapping a vast catalog of software tools that were already in circulation, and he felt certain that the same catalog could empower Aurora's designers. The key, he observed, was the use of Windows NT, not as a poor man's UNIX, but as a unique operating system with its own set of benefits — like the ability to use Excel spreadsheets for programming.

Armed with the voluminous catalog of Windows goodies, software designer Dan Proskauer went on a shopping spree. Was the Excel cell too limiting? Buy a program to enhance it. Need a good user interface? Buy it. Then fine-tune everything to the peculiarities of semiconductor test programming.

Windows NT would prove d'Arbeloff's instincts right. The operating system might not be up to UNIX for heavy-duty computing, but it had its own bag of tricks that more than compensated for any shortcomings. Besides, Excel would give the new test system a huge competitive advantage: Every potential customer already had in its employ legions of engineers who were already familiar with it and could thus learn to program the new system in no time.

To design the special chip needed for the compensation scheme, Teradyne turned to an engineer named Rick Meuthing. D'Arbeloff had come across Meuthing at a trade show more than two years earlier. Meuthing, who had worked for ASIX (later acquired by Credence Systems), knew his way around the world of temperature-compensated CMOS, and

[1]GeoTel and Vertigo

the trade-show conversation led to an assignment to design a chip that could serve as the heart of a new test system, while steering clear of Credence's patents. (At this point, of course, the whole Aurora project had yet to be conceived. D'Arbeloff was simply buying ammunition.) In 1994 Meuthing collaborated with Hap Walker on the design of a test chip, and when Aurora was born in early 1995, Walker's final version of this chip became the centerpiece of the new venture. It went into pilot fabrication in June, 1995.

In 1996, Motorola chose Hewlett-Packard to supply test systems for its low-end (8-bit) microcontrollers. This was unsettling news, for several reasons. First, it was another in a series of signs that the titan of the test instrument world was (again) becoming serious about semiconductor testing. Second, as mentioned, Motorola was a valued Teradyne customer, and Teradyne's absence from the "final four" was a blow. Third, although Teradyne retained the franchise to test Motorola's larger (16-bit) microcontrollers with its J971 and J973, the low end was where the volume was, and there was the haunting specter of hundreds of H-P testers filling the test floor.

But the excellent relations that had always marked the Teradyne-Motorola relationship were unruffled, and the semiconductor maker eagerly accepted Teradyne's invitation to review the work in progress at Aurora. Aurora also offered three other chip makers the chance to kibitz and to contribute their suggestions: IBM, Microchip, and VLSI Technology.

Work intensified on the new tester throughout 1996. A trademark search revealed possible problems with the name Aurora, and the group was renamed Integra. The product itself was given the designator J750. Several chip makers shied away from the task of building the needed custom gate array because of its complex analog cell, but in September Analog Devices agreed to tackle it, and Taiwan Semiconductor (TSMC) signed on as the foundry.

In June, 1997, Integra marketer Doug Elder received the first order, for five systems, from Microchip. The first system was shipped in January, 1998, and orders continued to trickle in until the fourth quarter, when the dam burst and the J750 was suddenly the hottest product in the ATE industry. In the coming years, hundreds of testers would be shipped, and three years after shipping its first product, Integra would be a $200-million-plus operation. The J750 technology, moreover, would migrate to other Teradyne divisions, enabling them to expand their product lines and enter previously untapped markets.

The Integra story again proved d'Arbeloff's ability to sense an opening, devise a strategy, marshal the necessary talent, and execute brilliantly. The key in this case was the determination to enter a new market despite the indifference of the existing organization. The creation of Aurora/Integra was a bold stroke that could have ended in failure — like d'Arbeloff's earlier failure with the J401. But this time it all clicked, and the J750 boasted the fastest takeoff of any Teradyne product, ever. Perhaps most satisfying of all was the decision by Motorola, in 1998, to rethink its choice of testers for its 8-bit microcontroller line and to give the business to Teradyne's Integra Division.

Kinetrix

Even before the first Integra tester hit the market, d'Arbeloff was off on another expedition into territory new to Teradyne. This time it was the world of device handlers. Virtually all ATE production-line test systems are coupled to handlers or probers, which serve to provide the mechanical link between the test electronics and the chip under test. The combination of tester and handler or prober is often considered as a single unit, and tester makers devote considerable effort to the task of creating a seamless interface between the two units.

The Japanese tester giant, Advantest, produced its own line of handlers, but otherwise handlers and probers were the products of a dozen or so smaller, independent companies, notably including Daymarc, the Massachusetts-based company founded by d'Arbeloff's old friend Nick Cedrone. Rumors had been floated from time to time about a Teradyne-Daymarc union, but, although the two companies routinely joined forces in serving customers, d'Arbeloff saw little attraction in such an acquisition. At any rate, Daymarc's purchase by Cohu, Inc. ended any such speculation.

It was not the handler business per se that put d'Arbeloff off, but his aversion to entering any new market simply for "a piece of the action." If he were ever to enter the handler or prober business, it would have to be with a major advance in the technology.

The prospect of such an advance grew out of d'Arbeloff's close association with his alma mater, MIT. There he lectured on entrepreneurship and served on a number of committees, including the Institute's Board of Trustees and an advisory group assigned to the Mechanical Engineering Department. These activities brought him into contact with Alex Slocum, a brilliant and innovative member of MIT's Mechanical Engineering faculty. One conversation led to another, and d'Arbeloff was soon convinced that Slocum's genius was exactly the magic carpet on which Teradyne might ride triumphantly into the device handler business — if not the entire "back end" of semiconductor production.

The central idea was to create a line of handling equipment whose reliability would be an order of magnitude greater than anything currently available (an obvious echo of Teradyne's original business plan). Since Teradyne had no installed base, it was not bound by legacy and could start life with a clean slate. The first products would be a handler and sorter based on a new concept: testing devices loaded in trays, rather than individually. The batch handling would dramatically improve throughput and eliminate most of the jams plaguing conventional handlers. Each tray would be tested at temperature and later sorted off-line, using bar-code identification to keep everything straight.

Others had tinkered with this concept, but no one was shipping working equipment. Most important, the Teradyne handler would vastly improve every aspect of the technology. To ensure optimal air flow around the tray, a special wind tunnel was rigged up. To ensure structural integrity, the steel chassis was build like a tank. Trays were raised and lowered silently and smoothly by corkscrew steel shafts. An ingenious new contactor design promised major improvement in reliability.

The new venture, called Kinetrix, opened its doors in Bedford, New Hampshire. The chief technologist was Alex Slocum (who remained on the MIT faculty while visiting Kinetrix regularly), and Slocum's brother Rick managed the operation. Before long, the unit was staffed by dozens of engineers, and the new handler (Galileo) and prober (Polaris) began taking shape.

As events played out, Kinetrix would not become another Integra, and in fact a slimmed-down Kinetrix would eventually be absorbed by Integra and move from Bedford, New Hampshire to Integra's plant in Bedford, Massachusetts. What had gone wrong? The test-in-tray concept, so attractive at first glance, encountered resistance in the marketplace. A technology shift occurred, and attention was diverted to strip handlers. And, of course, the existing handler industry did not roll over and play dead when Teradyne entered its turf, but created its own tray and strip handlers. Without question, Kinetrix gave Teradyne valuable experience in a technology of vital importance to any tester maker — experience that may yet yield a dividend.

Megatest

Early in 1995, Teradyne began exploratory talks with Megatest Corporation, an ATE maker based in San Jose. The Company, 20 years old and publicly owned, had created a stir years before with the introduction of the Mega One VLSI test system, which featured a "tester-per-pin" architecture that quickly became the talk of the industry. For cost considerations, other testers, including Teradyne's early VLSI systems, had been based on a "shared resource" architecture, in which the system's timing and formatting circuits are shared by the system's drivers and detectors. The Mega One, by dedicating a driver and detector to each timing and formatting circuit, greatly increased system accuracy and flexibility, albeit at higher cost than the shared-resource approach. Megatest promoted its innovation aggressively, and some semiconductor makers, including Texas Instruments, opted for the new system. In addition, Megatest produced a line of flash memory testers, many of which had been sold to Intel. All told, Megatest sold $100 million worth of semiconductor test equipment in 1994.

Some thought the acquisition made sense; others did not. Those in favor made several strong arguments. Megatest's installed base at Texas Instruments could be exploited. The flash memory tester line would make Teradyne the largest factor in a business likely to grow rapidly. The Megatest engineering team represented an important talent pool, in an era when experienced test system developers were hard to find. Finally, the deal might make sense as a way of keeping the Megatest resources out of the hands of competitors. In the end the "for" arguments carried the day, and Megatest was acquired in December, 1994, at a price of 6.8 million shares, valued, at the time of the announcement, at about $265 million.

The director and officer corps saw some realignment in 1995. Franklin "Pitch" Johnson, the California venture capitalist, resigned from the Board, and James Bagley, Vice Chairman of semiconductor equipment giant Applied Materials, replaced him. (Bagley was well known to d'Arbeloff and to Teradyne's semiconductor test sales people; years earlier, as a Texas Instruments manager, he was Teradyne's *bête noire* as it competed against TI's in-house test group.) As Chamillard became President and Chief Operating Officer, Jim Prestridge and Owen Robbins were named Vice Chairmen. With the Megatest acquisition, that Company's former President, Jack Halter, became a Teradyne Vice President. With my retirement, George d'Arbeloff became the Company's Vice President for Corporate Relations. The younger d'Arbeloff had been Vice President of the Semiconductor Test Group, a layer of management now seen as one layer too many.

Now it was only a matter of time before Chamillard became CEO as well as President, but until the actual event the Company would be led by d'Arbeloff (de jure) and Chamillard (de facto). Under the best of circumstances such a transition has its awkward moments. If there were any in this case, they were not obvious to the onlookers. D'Arbeloff gradually ceded more and more of the key decisions to his successor-to-be, and Chamillard, who must have been anxious to get on with it, took only as much authority as his mentor was willing to yield at any given point.

After a four-year stretch that saw sales double and net income grow sevenfold, Teradyne braced for the bust that usually followed such booms. And indeed, sales did decline in 1996 (by less than two percent), and while profits predictably sank, there was still more than $90 million dollars of after tax income — actually $125 million before a one-time charge for the consolidation of the Megatest product line. The damage was much greater among Teradyne's competitors, some of whom fell deeply into the red.

Salvation came to Teradyne, as it so often had in the past, on the wings of diversification. Board test sales grew by 20 percent, far outpacing market growth, and Connection Systems tacked on another 15 percent. The Telecommunications Division wrapped up the Deutsche Telekom installation and began aggressively marketing 4TEL II, and the software test units doubled their sales (from a very small base, of course). Even in the semiconductor test sector, the Company was able to moderate the effects of the downturn, thanks to a large backlog. All in all, it was hard to get too depressed over the slight interruption in over-all sales growth, especially against the background of the heaviest outpouring of new products in the Company's history.

Of these, a new mixed-signal test system, called Catalyst, promised to make the biggest waves. With the A500 product line still going strong, Dave Sulman and his Industrial/ Consumer Division moved smartly to stake claim to an exciting new market, the testing of a new breed of integrated circuit known as a "System on a Chip." Catalyst also caught another new wave: wireless. As chip makers scrambled to achieve the high-frequency test performance required by cellular phones and other wireless devices, Catalyst was ready, boasting performance "from dc to light."

In the digital area, the first order of business was the integration of the Megatest engineering staff into Teradyne west-coast facilities. The solution was straightforward in concept, difficult in execution. The memory resources of Teradyne and Megatest were combined into a single Memory Test Division (MTD), based in Agoura Hills under Tom Felding, a Teradyne veteran. Similarly, the VLSI logic resources were pooled to form a VLSI Test Division (VTD), also in Agoura Hills, under Wayne Ponik, also a long-time Teradyner. Megatest's San Jose plant was placed within MTD, and another Megatest engineering group in Fridley, Minnesota was assigned to the new VTD unit.

All this realignment failed to slow down the outpouring of new products. In VLSI logic, the J971, by now something of a standard with 300 systems on line, was succeeded by the J973, with twice the channels and twice the speed of its predecessor, at only a modest increase in price. The J973 was now entering the speed domain (400/800 MHz) where water-cooling was necessary, and one could almost hear the groans from Nick "Speed Kills" DeWolf. In memory test, a new 64-in-parallel tester, called Marlin,[1] came out of the gate late in the year with a spate of orders from makers of 64-megabit DRAMs.

In board test, the kind of rampant success that had eluded the group for so many years finally struck. In the military market, the B-2 test depot contract was completed, and the pro-

[1]The christening of new systems with names (Catalyst, Marlin) offers an interesting sidelight in the history of Teradyne marketing. Nick DeWolf had been fond of trade names (ACT I, the SLOT Machine), but since his departure product names had stayed within Teradyne's alphanumeric system (with the notable exception of 4TEL). But acquisitions had brought new promotional thinking. The Walnut Creek unit (ex-Zehntel) introduced a new board tester called Spectrum. Hammer came with Hammer, and the Systems and Software Test unit was selling TestMaster. The old order was breaking down, and names rather than type numbers were suddenly in vogue. But not everywhere: The VLSI test division at first called its new J973 system "Magnum," then retreated in fear that someone might associate the word with guns.

gram's prime contractor, Northrop Grumman, agreed to join Teradyne in installing its digital test units in another major military program for the U.S. Army. Teradyne's clear sweep of the military market for digital test systems prompted the editor of a leading industry newsletter[1] to cite "Teradyne's success in capturing the digital portion of virtually every military ATE test system being built or upgraded today," adding that "They have put themselves in the driver's seat for military digital test for years to come."

At Assembly Test Boston (ATB), however, Division Manager Jeff Hotchkiss was leaving the driver's seat to join the new Chamillard team as director of planning. He was succeeded by an outsider with strong military contract experience. The new man, Bob Feingold, was quickly made a Vice President of the Company.

The Walnut Creek half of the board test organization ran its total of Z1800 shipments to 2000, and it equipped the system with new options as fast as its engineers could dream them up. The Z1800, which had started life as a bare-bones in-circuit tester, was now a Christmas tree on which to hang boundary-scan capability, vectorless programming, an analog digital signal processing option (PRISM-Z), a mechanical handler (from Okano), and a catalog full of other options. Then, with the low end of the board test market firmly in hand, Walnut Creek began moving upscale with the introduction of Spectrum, industry's first VXI-based in-circuit tester, boasting Windows NT programming and a long list of analog and digital test capabilities.

Walnut Creek was thus becoming a happy exception to d'Arbeloff's dictum about acquisitions. To be sure, Teradyne still trailed Hewlett-Packard and GenRad in the total board test market, but with board test sales of more than $150 mil-

[1]Bode, Fred, *VXIbus Newsletter*, October/November 1996

lion — and growing — the prospects were the brightest they had been in years. Thus encouraged, the Company decided to expand its product reach by moving into the optical board inspection market, through the acquisition of a small New Jersey company, Control Automation, Inc. Control Automation was already in the business, with an installed base of 450 systems and with some talented engineers, who were soon moved across the country to Walnut Creek.

Connection Systems, with sales now approaching $200 million, was parlaying its HDM line into a commanding position in the high-density connector market. It now numbered some of the highest flying of all electronics manufacturers — Sun Microsystems, Cisco, EMC — among its leading customers, and it raced to keep up with demand, adding capacity in Nashua and Dallas. And TCS opened a second front by licensing established connector maker Molex as a second source for HDM.

The TCS product genealogy is as easy to delineate as that of Teradyne's IC test systems. Its high-density MiniMax connector was the forerunner of the HD+ connector, which then spawned a long list of derivatives: UHD (chiefly for the military market), HDM (High Density Metric), HDM+ (shielded HDM), VHDM (very HDM), etc. These were all press-fit connectors, in which the male connector is press-fit into the mating female part. The one excursion to date outside the press-fit envelope was an innovative design called Beam On Pad, intended for use in surface-mount applications. Teradyne put the connector to good use in its VLSI test systems, but the rest of the world was not yet ready for such a radical departure from connector tradition.

D'Arbeloff, meanwhile, was still at work building his projected software test empire. This time the target was Midnight

Networks, Inc., a four-year-old, Waltham-based maker of equipment to test communications networks, including elements of the Internet. Midnight fielded two product lines, ANVL to test protocol interactions and Avalanche to simulate and test complex networks.

In 1996 Teradyne's Board of Directors gained its first female member. The newcomer was Patricia Wolpert, General Manager of IBM's Northeastern Area. A year later, the Company would name its first female Vice President.

22

• • •

Chamillard

Teradyne's first two presidents were children of privilege, blessed with distinguished ancestors, well-off parents, governesses, and all the other advantages of good breeding. They also graduated from MIT just as the electronics industry was poised to explode, giving them and their classmates a ticket of admission to the greatest show on earth. Neither had to climb the corporate ladder at Teradyne; they began standing on the top rungs.

This is not to detract from their achievements, which are monumental, but only to illustrate the contrast between their backgrounds and that of Teradyne's third president.

George Chamillard attended public schools in Boston's Roxbury section, graduated from Norton (MA) High School, and graduated from Boston's Wentworth Institute with an Associate's Degree. Later he picked up BS and MBA degrees from Northeastern University. His 30-year climb to the presidency of Teradyne began with a job as a test engineer.

Chamillard is also the first President of this Boston Company who is a bona fide Bostonian. DeWolf and d'Arbeloff, along with Prestridge, Robbins, and most of the other senior executives, converged on Boston from Paris and Florida and Oklahoma and Philadelphia and wherever, but Chamillard spent his childhood within a few miles of Harrison Avenue.

Considering the legions of intelligent, aggressive, well-credentialed, eminently successful people who have passed through the Teradyne ranks over the years, the ascension of George Chamillard to the top is remarkable.

Or maybe not so remarkable at all.

George Chamillard was born on December 28, 1938, the second of three sons of Albert and Helen Chamillard. His father was a chemical engineer and the first member of the family to earn a college degree (from Northeastern University). The Chamillard family lived in the Roxbury section of Boston, and George attended public elementary schools there and entered Boston English High School, in Boston's South End.

In his sophomore year the Chamillard family moved to Norton, a small farm town in southeastern Massachusetts, where Chamillard's forebears had settled after arriving from France by way of Canada. George graduated from Norton High School in 1956. Then, with no particular objective in sight, he entered Wentworth Institute, specializing in Industrial Electronics. Wentworth's Boston location and its focus on the practical side of electronics would make it a natural feeder for Teradyne, and several members of Wentworth's Class of 1958 would eventually spend long, productive careers with the Company.

In 1958, however, Chamillard was headed for the open sea. In his senior year at Norton High, he had joined the U.S. Coast Guard Reserve, and he had since faithfully attended his weekly meetings. Now, upon graduation from Wentworth, his Coast Guard status offered what seemed to be an attractive opportunity: a six-month course in electronics at the base in Groton,

Connecticut. He tested for admission, scored well, and was soon in Groton. Six months later he was bound for sea duty on various Coast Guard ships. He was also a married man.

Norton is not far from the Town of Foxboro, home of the Foxboro Company. Chamillard had worked summers at the Foxboro Company while in school, and he naturally gravitated to Foxboro on separation from the Coast Guard, landing a position in research. Months of winding toroids and measuring cores dulled his appetite for research, and he shifted into product engineering, eventually heading product, test, and factory engineering at Foxboro's Bridgewater plant. By now he was convinced that his heart was in the manufacturing side of electronics.

While working at Foxboro, Chamillard began an eight-year stretch of night-school classes at Northeastern University, traveling three times a week (two weeknights plus Saturday) from his home in Attleboro north to Northeastern's Boston campus. A mediocre student in high school, he bore down at Northeastern, and graduated high in his class with a BS degree in Industrial Technology. Then more night school and an MBA, also from Northeastern.

It is not known whether The Foxboro Company realized what a prize it had in this young manufacturing engineer, slogging through eight years of night school while raising a family and doing well at his "day job" in its factory, but this was exactly the kind of young man Teradyne was looking for in the mid-1960s. The connection came via Wentworth classmates who had joined the new company and were enthusiastic boosters. Chamillard turned down one Teradyne offer in 1967, but in 1969, while picking up a classmate in Teradyne's Essex Street lobby, he encountered Alex d'Arbeloff ("it was probably a setup," he says now), and Teradyne's master recruiter quickly snagged his quarry.

Like most new hires, Chamillard spent his first six months in Teradyne's test department. There followed a number of assignments in Manufacturing and Operations until, in 1975, he was transferred to Teradyne's Digital Test Division, in Chatsworth, California to help Jack Salvador manage operations.

Many of the mileposts in Chamillard's Teradyne career have already been noted, but one contribution is worth special attention: the decision to pool the resources of Teradyne's various circuit-board facilities into a single "foundry." (In time, there would be two foundries, one in Massachusetts, another in California.)

By creating a centralized foundry, he argued, Teradyne could afford the best production equipment available and begin to exploit the economies of scale that had been so elusive. Even more important, the foundries would be large enough and the challenges interesting enough to attract some of the Company's best talent. Another payoff, he noted, would be the development of much-needed manufacturing standards, since automated lines would not permit engineers to design by whim. These ideas resonated with Owen Robbins, who became a strong advocate of the foundry concept as one more weapon in his campaign to turn manufacturing efficiency into a competitive edge.

All this, of course, was many years before the advent of what is now called the electronic manufacturing services (EMS) industry and the semiconductor foundries that now dot the Taiwan landscape. In 1980, the term "electronics foundry" sounded to many like an oxymoron, and many Teradyne employees must have found it difficult explaining to their friends why they were doing working in a foundry. But the term was just right. Chamillard thought so, Robbins thought so, and Nick DeWolf, had he been asked, would have been ecstatic.

Chamillard's promise was spotted early (see page 152), but his "breakthrough role" was his tenure running Teradyne Connection Systems. To outsiders, it may have seemed strange that a Company known primarily for its ATE would choose a president from its connector ranks, but of course Chamillard was an old ATE hand, and the tour at TCS gave him a perspective on Teradyne's *two* major businesses that no one else could match. So, when his appointment as Executive Vice President effectively named him the "designated successor," no one inside the Company was terribly surprised.

Chamillard's wife Maureen is an Irish colleen, and he maintains a home away from home in Galway as well as houses in Massachusetts and New Hampshire. He has followed his predecessor onto the Board and Chairmanship of the Massachusetts High Technology Council. In his relatively short time at Teradyne's helm he has become a highly visible member of the high-tech community. He almost certainly does not welcome celebrity, but it goes with the territory.

Chamillard is not a flashy manager. Neither is d'Arbeloff, for that matter, but d'Arbeloff does have a certain continental panache, while Chamillard is essentially the stolid New Englander, not unlike the majority of workers surrounding him at Harrison Avenue. Like the first Teradyne testers, he makes it on the basis of capability, integrity, and quality.

23

THE CHAMILLARD ERA BEGINS

Although the process of transition was spread over several years, it is 1997 that stands as the year in which the d'Arbeloff era at Teradyne ended. The founder would continue to serve on the Board of Directors for a few more years, sharing his strategic visions and overseeing his cherished software-test group, but by the end of 1997 Teradyne would essentially be George Chamillard's Company, as he became CEO as well as President. The hand-over took place with Teradyne at the top of its game. The mild downturn of 1996 was a thing of the past, and in 1997 the Company posted record sales ($1.266 billion) and record net profit ($127.6 million).

The recovery accelerated throughout the year. Bookings, which bottomed at $215 million in the third quarter of 1996, reached $461 million in the third quarter of 1997 and held that level in the fourth quarter. The new mixed-signal test system, Catalyst, was an instant winner, and by year-end it represented about 30 percent of ICD's mixed-signal business, even though the older A500-series itself posted record sales. Similarly, the Memory Test Division's Marlin tester and the VLSI Test Division's J973 were instant successes. The Agoura Hills operation, once again running out of space, acquired two buildings near its present sites and moved production of the J921 VLSI Test System to the Japan Division's plant in Kumamoto, Japan.

In board test, sales could not equal a 1996 level inflated by the B-2 contract, but there was still much to celebrate, including the booking of 100 new Spectrum testers, the capture of a large Marine Corps contract, and the introduction of a new "Flying Prober" tester, made to Teradyne's specifications by Japan's Shindenshi Corporation. The Telecommunications Division received a GTE order for the system-wide deployment of 4TEL II, plus two $8 million 4TEL orders, from telephone companies in Argentina and Hungary.

Not to be outdone by its ATE sister divisions, Connection Systems ratcheted sales up 23 percent, to well over $200 million. Building on the success of the HDM connector, TCS introduced the VHDM, an eight-row design that offered a density of 100 contacts per inch. The TCS printed-circuit facility was again expanded, and the Division began moving into the 78,000-square-foot ex-Harris site just down the road from its Simon Street headquarters.

Space was tight at Teradyne's Boston headquarters, despite the leasing of various sites in the City, and of course the parking issue was a never-ending headache. The Company, which had scraped by in Boston for so many years through opportunistic real-estate investments, had played out that string. The next step, inevitably, would be a more traditional campus in the suburbs. After much searching and polling of employees, the Company purchased a 45-acre tract in North Reading, Massachusetts, not far from the junction of Interstate Routes 95 (Mass. Route 128) and 93.

Underscoring the ending of the d'Arbeloff era, Jim Prestridge and Owen Robbins both decided to retire in 1997. Though both would remain on the Board, along with d'Arbeloff, Teradyne's "big three" were suddenly removed from the day-to-day operations of the Company. But d'Arbeloff was not ready to cut the cords, and he enlisted the aid of both Prestridge and Robbins in directing the affairs of the software test group.

The new Teradyne management team viewed the thrust into software testing with ambivalence, seeing it as a possible distraction, but with d'Arbeloff, Prestridge, and Robbins masterminding the effort, no one was ready to write it off. Then, in late 1997, d'Arbeloff added two more Boston-area horses to his software test stable: Softbridge, a Cambridge company specializing in the testing of enterprise software used by financial service companies, and RSW, a Watertown-based provider of Web-testing software, named for its three founders, Yuri Rappaport, Armand Sundaram, and Ray Wach. The list of new software test ventures now numbered five: SST, Hammer, Midnight, RSW, and Softbridge.

As if he needed additional claims on his time, d'Arbeloff in 1997 accepted MIT's invitation to become Chairman of its Board of Trustees. The Institute's decision to reach into industry rather than academia in making this important appointment was, among other things, a sign of its intention to become more closely coupled to the world of industry d'Arbeloff represented. In another sense, however, the appointment was simply the culmination of d'Arbeloff's long, close relationship with MIT — a relationship that served as a conduit through which graduates and sometimes faculty members moved from MIT to Teradyne. Ever on the prowl for talent, d'Arbeloff would continue to prospect for human gold, but now from an even more advantageous position.

The Chamillard team seemed to be set in place at the end of 1997. The Chief Financial Officer was Jeff Hotchkiss, whose long career at Teradyne included stops in sales, marketing, division management, and planning. Three Vice Presidents — Mike Bradley, Ed Rogas, and John Casey — were given broadly defined responsibilities involving semiconductor test sales, marketing, operations, and customer service. Tom Newman, the long-time Customer Service manager, was installed as Vice President for Corporate Relations, replacing George d'Arbeloff.

The youngest of the three d'Arbeloff brothers had contributed much to Teradyne over three decades, working in software development under Milt Collins, in Tokyo when Teradyne was new on the Japanese scene, as ATE Field Service Manager, as Manager of the Industrial/Consumer Group, and as head of the Semiconductor Test Group. Few people played as many or as widely varied roles in the history of the Company. But he was always in the shadows cast by his older brothers, both CEOs of important technology companies, and, when the chance came to join an entrepreneurial venture in the computer field, he left Teradyne in search of a new set of challenges.

Randy Stone, Teradyne's first woman Vice President, slipped into Newman's position as Vice President for Global Customer Service, presiding over a massive world organization responsible for system installation, repair, maintenance training, and the management of almost $200 million worth of spare parts at 11 sites around the world.

A number of other new vice presidents were named in 1997: Ron Butler, who had managed the Walnut Creek board test unit and the post-acquisition Megatest, became Vice

President for Human Resources. George Cary, a 27-year veteran manufacturing manager, took over responsibility for Teradyne's east- and west-coast circuit-board foundries. Tom Felding, manager of the Memory Test Division, and Wayne Ponik, manager of the VLSI Test Division, were both named Vice Presidents. A new hire, Wayne Lasson, succeeded John Casey as manager of the Telecommunications Division, and ICD marketer Mike Luttati became Vice President for North American semiconductor test sales.

Shortly after year-end, a new General Counsel named Jack VanWoerkom was hired and named a Vice President. His tenure would be relatively brief, as would that of his successor. With a few exceptions, Teradyne would have far greater success in promoting from within than it would have with executives entering the Company at a senior level.

As the Teradyne of Alex d'Arbeloff morphed into the Teradyne of George Chamillard, no one outside the Company would have noticed any change in direction or philosophy or culture. Chamillard was now the one moving the pieces around the board, but the chessmen were mostly old friends and colleagues, thoroughly steeped in both the old DeWolf-d'Arbeloff culture and the TQM religion of the 90s.

Also familiar with the Teradyne culture was new Director Vincent O'Reilly, formerly a Vice Chairman of Coopers & Lybrand and now a Senior Lecturer at Boston College. Coopers was Teradyne's auditor for more than a quarter century, a number of Teradyne's key managers were Coopers' alumni, and "Vin" himself had been a frequent visitor to Teradyne for many years.

Throughout most of the nineties, Teradyne's growth could be largely attributed to the semiconductor industry's healthy appetite for capital equipment. Teradyne had gained market share, to be sure, but the rising tide had helped lift all boats. The real test came in 1998, when the tide ran out. When industry analysts finally crunched their market numbers, they came up with the grim fact that orders for semiconductor capital equipment had plunged by about 70 percent during the year. It was, in short, the perfect background against which to test Teradyne's thesis that TQM was making a difference.

The results were eye-popping. In this dreadful year — the first full year under its new CEO — Teradyne lifted its sales to a record $1.5 billion. In a year when other ATE companies were drowning in red ink, the Company earned over $100 million after taxes. Its semiconductor test sales grew by 14 percent, its connection system sales by 19 percent. In the former, shipments were sheltered by a large starting backlog. In the latter, the growth was exactly what it appeared, as TCS rode the Internet tide, supplying some of the fastest growing makers of servers and routers, as well as the data storage market. To serve its largest Silicon Valley customer, TCS opened a new plant in Milpitas, California, and in Nashua, the Company invested $11 million in another major expansion of its printed-circuit facility.

TCS also saw turnover at the top in 1998, as Ron Dias returned to Boston to head ATE Operations. The new manager was Rick Schneider, an 11-year veteran at TCS. Schneider, an aggressive, analytical manager, had come to TCS from Texas Instruments in 1987 and had served as the Division's Marketing Manager for several years. Under Dias, TCS had grown impressively, and it was a tough act for Schneider to follow. No one could possibly have foreseen that in Schneider's first two years on the job, TCS sales would almost triple.

Teradyne's relatively strong performance in a desperately bad year produced outsized gains in market share. The situation was especially bright in mixed-signal test, the fastest growing of all ATE sectors these days, where Catalyst and the A500 combined to give Teradyne a vise-like grip on the market. LTX was heavily promoting its Fusion system as a system-on-a-chip tester, but Catalyst was winning the lion's share of the SOC orders. Hewlett-Packard and Credence were also active in the market, but with its rapidly growing Catalyst base Teradyne was building a fortress difficult to attack.

In logic testing, the new star in the heavens was Teradyne's Integra J750, which scored a direct hit on the microcontroller test market. The first J750 was shipped (to Microchip) in the second quarter, and by year-end Integra had 50 systems from 12 customers in its order book. It was clear that the architecture was ideal for the testing of low-margin, commodity-type components, and options extending the system's reach to memories and analog circuits were speedily developed by Integra's engineering staff, while in California, Teradyne's Memory Test Division began planning a flash memory tester using the same platform.

The Marlin memory tester was enjoying great success as a DRAM tester in wafer-probe applications, with almost 100 systems shipped by the end of 1997. A new memory tester, called Aries, also made its debut in 1998. Aries was developed in response to a growing clamor surrounding the Rambus architecture, a scheme for effectively increasing a memory's speed. In Aries Teradyne was sure it had the best solution to a difficult testing problem, but the Rambus architecture proved controversial, and the on-again, off-again demand for Rambus parts was a problem.

Trying to "psyche out" a confusing market was also a continuing challenge at the Telecommunications Division. It was clear to everyone that telephone operating companies, in order to survive in the new digital era, would have to increase the effective bandwidth of their twisted pairs of copper lines, using a signal compression scheme known as DSL (Digital Subscriber Loop). It was equally clear that it would be necessary to test each existing line to see whether it was DSL-ready. Finally, all eyes were on the "last mile" of wire leading to each subscriber's premises, because it was generally understood that this was the crucial competitive battleground. Teradyne's 4TEL system was installed on more than 100 million "last miles," and to seize the moment the Company adapted the system to enable it to prequalify phone lines for DSL capability. If the telephone companies embarked on a crash DSL program, Teradyne would be ready, just as its memory test division was ready for Rambus.

In the event, neither Rambus nor DSL would trigger a crash program, but that is almost beside the point. Ten or 20 years before, Teradyne might have been late, rather than early, to market. (Recall Teradyne's failure to produce a VLSI tester until Sentry had run the table.) Now Teradyne was in the enviable if somewhat frustrating position of being ahead of its customers.

Teradyne finally seemed to be finding its way in the board test market, growing 25 percent in a stagnant marketplace. The growth was evenly spread across the Company's commercial and military lines, with the Spectrum in-circuit tester and the dependable CASS business both contributing. In addition, the Company was suddenly a factor in the optical inspection market, off the sale of more than 40 AOI (automated optical inspection) systems to Intel. The product was a vastly improved version of the AOI system that arrived at Teradyne with the acquisition of Control Automation, in 1996.

Since the acquisition of Zehntel in 1987, Teradyne had maintained two separate board test operations: the Walnut Creek Division and Assembly Test Boston, serving the commercial and military markets, respectively. In 1998 the two were combined into a new Assembly Test Division under Joe Wrinn, a Teradyne engineer who had been running the Walnut Creek unit.

Some alert readers may have noticed a certain organizational oscillation at work here. The semiconductor test divisions, east and west coast, begin as separate units, then are merged into a single group; then, years later, they are unmerged into separate divisions (and, in the fullness of time, will be re-merged). The board test divisions are merged, then unmerged, then re-merged. Each fusion or diffusion is justified, in memos to employees and press releases, on the basis of revealed truth: The move to a single entity (or two entities, take your choice) is obviously the only logical way to proceed.

There is a less cynical spin: The organization at Teradyne was always made to fit the talent at hand, as well as the marketing situation at the time. This was as it should have been. Teradyne people weren't hammered into slots; they made the slots.

The board test organization also gained a new sales and support manager in 1998, as Peter Kelley, who had steered TCS operations in Japan and Europe for a decade, decided to move into ATE territory.

Wrinn became a Vice President with the board-test reorganization, and the Company also named three other new vice presidents: Rick Schneider, managing TCS; Roger Saylor, a 20-year Teradyne employee directing semiconductor test sales in

Southeast Asia; and a newly hired Information Services Manager, Rob Strickland. Strickland would soon leave, but Schneider was just two years away from running Teradyne's largest division, and Saylor would soon become worldwide Sales Manager for semiconductor test products.

The patchwork quilt of software test companies put together by d'Arbeloff now begged for rationalization. A few stabs were made: The TestMaster product out of the Systems and Software Test unit was adapted for use with the products of Hammer, and there was a similar collaboration between SST and RSW. A new effort, called NorthStar, pooled resources of Midnight and the Telecommunications Division. But in general, Teradyne had on its hands five small, separate operations, each going its own way, linked together only by the oversight of d'Arbeloff, Prestridge, and Robbins, plus some G&A services provided by the parent Company. The five, as might be expected, included both winners and losers. Hammer was by far the largest and most successful. RSW's sales of Web-testing software were soaring. Midnight and Softbridge were stumbling, and SST was stable at a low level of sales.

Chamillard shared d'Arbeloff's view of the software test group as a major opportunity, but he was less sure that Teradyne was the right sponsor. In 1995, the decision to enter software testing had been based on a set of assumptions: First, that the venture would be synergistic, as Teradyne's engineers would apply and validate new software test capabilities. Second, that it would bring Teradyne into a new and potentially explosive market. Third, that it would serve as a hedge against unpredictable hardware sales. Fourth, that the Company could easily afford to assign some of its senior management talent to the venture.

It was now becoming clear that most of these assumptions were invalid. There was virtually no synergy. Teradyne's core hardware test business was itself exploding and consuming all its human resources. It was also obvious by now that the software test business was worlds apart from the capital equipment business, and much of the expertise accumulated by Teradyne over 40 years was of little value to the new ventures. Teradyne sold IC testers by laboriously building relationships with chip makers over many years. RSW sold packaged software in minutes, over the Web.

The software test units were also becoming restless. In 1998 the stock market was routinely valuing start-ups at 100 times earnings or more, and to many engineers in the software test units, fame and fortune were only an IPO away. Not that the Teradyne stock they held was doing badly, but the real payoff could come only via a spin-off.

In 1999 the Company took a dramatic step to accelerate the software group's development, when it named Jeff Hotchkiss as its new manager. Now the die was cast. Hotchkiss was clearly one of Teradyne's most valuable executives, and his departure from Harrison Avenue was a signal that Teradyne was serious about melding the various software test units into a cohesive entity. He was playing a good hand; sales at the software test group grew by 70 percent in 1999, and hundreds of blue-chip customers, from a cross-section of industry and commerce, had come to rely on the Hammer and RSW products.

To replace Hotchkiss as CFO, Teradyne chose Mike Bradley. The mild-mannered Bradley had a distinguished career at Teradyne, most notably as the Product Manager of the A500, as the Company's first TQM Manager, and as world-wide sales manager for semiconductor test products. He had also been a member of Teradyne's Executive Committee since 1996.

To succeed Bradley as semiconductor test sales manager, Chamillard chose Roger Saylor. Saylor, like Prestridge and Rogas an Annapolis graduate, had spent his Teradyne years in a variety of sales and marketing positions and had recently managed Southeast Asia Sales from a base in Singapore.

Over-all, sales in 1999 grew by a solid 20 percent, to just under $1.8 billion, while net income almost doubled. Teradyne's grip on ATE world leadership tightened, as it funded a broad range of engineering programs, hired people by the thousands, and built new plants to house them. Despite stock buybacks and heavy investments to expand capacity, cash continued to mount, along with the market value of the Company's stock.

Just as Teradyne had shrugged off DeWolf's departure in 1971, it had apparently shrugged off d'Arbeloff's departure as CEO in 1997. Of course, no one knew better than George Chamillard that d'Arbeloff had placed in his hands a Company just hitting its true stride, a Company fine-tuned to the TQM pitch, and a Company with a deeply ingrained appetite for excellence. All that was left to do, as Chamillard explained facetiously to a reporter, was "not to screw up."

Semiconductor test sales climbed by 25 percent in 1999 (orders were up a breathtaking 130 percent), while Connection Systems grew by 39 percent, as its sales closed in on the $400 million mark. The technology behind the HDM and VHDM connectors was given another spin with the introduction of VHDM-HSD (high-speed differential), and a new stacking connector, called NexLev, was added to the line. NexLev was remarkable as a thrust outside TCS's traditional turf, as it was designed to interconnect circuit boards without benefit of a backplane.

Vice President Bush visits Teradyne, July 15, 1985.

The plant in Kumamoto, Japan.

Tom Felding and Mark Jagiela.

The original three-building complex at Agoura Hills, California.

Tom Newman.

Ron Dias.

John McCabe.

Randy Stone and Roger Saylor.

Jim Dawson and Ron Butler.

Marc Levine and Stu Osattin.

Don Hamman and Wayne Lasson.

The Integra J750.

Mike Bradley and Rick Schneider.

Ed Rogas.

Dave Sulman.

George Chamillard.

The North Reading campus.

John Petrolini, Joe Wrinn (standing) and Brad Robbins.

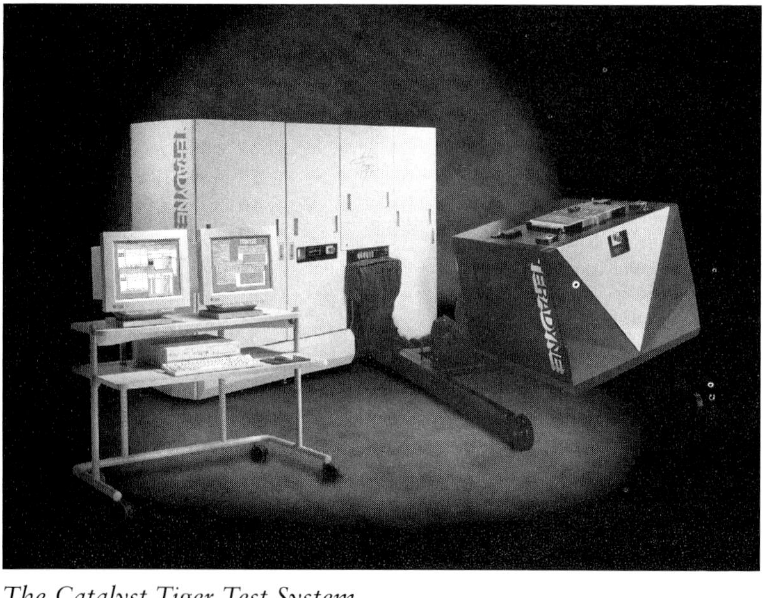

The Catalyst Tiger Test System.

The Catalyst test system thoroughly dominated the market for system-on-a-chip testing, with 500 systems booked and 400 shipped in its short lifetime. The J973 and J971 VLSI test systems feasted on the burgeoning market for microprocessors and microcontrollers, while the Integra J750 far exceeded the Company's wildest hopes, as it shipped the 100th unit in August, the 200th in December.

Sales were not particularly strong in the memory test, board test, or telecommunications (4TEL) areas, for varying reasons. The failure of telephone companies and their equipment makers to deploy DSL lines aggressively was a problem both for the Telecommunications Division and for the Assembly Test Division, whose latest version of its Spectrum board tester was aimed squarely at that market. Yet at all three of these divisions, important new products were taking shape in the engineering laboratories, and all three would make headlines the following year.

With orders pouring in from every direction, with the production lines churning out new products, with new plants opening everywhere, and with the cash hoard and the stock price both climbing, Chamillard could have been forgiven for wondering why people made such a fuss about managing a billion-dollar company. What could be easier?

But Chamillard had been through enough cycles to know that the growth rates now being experienced were unsustainable. Anyone with an ounce of sense must know that when semiconductor test orders grow 130 percent in a given year — and that from a reasonably healthy starting level — it was only a matter of time before you would be brought face-to-face with reality. It had happened before. It would happen again.

Yet there really was a difference this time. While past semiconductor booms had all fed on the consumer's appetite for personal computers stuffed with ICs, this time there was a gigantic new market promising to have an even more voracious appetite for chips. Two markets, actually. The Internet revolution was forcing the complete rebuilding of the world's communications networks, and the wireless revolution was soaking up vast quantities of mixed-signal chips to be tested on systems like the Catalyst. Each revolution spawned sub-revolutions and counter-revolutions, as companies scrambled for market position. And there was broad-line Teradyne, serving as a proxy for the whole lot.

One thing was sure: Teradyne was now a big company. Even allowing for the fact that a billion dollars was no longer the lofty threshold it once was, Teradyne in 1999 was still a far cry from the Teradyne of old. More than 7500 employees now worked for the Company, and the growth of recent years meant that most of them had not been around long enough to absorb the Teradyne culture. Old-timers who visited Teradyne's Harrison Avenue cafeteria found a sea of unfamiliar faces, and a reference to Alex d'Arbeloff might bring an "Alex who?" Only a small minority knew who Nick DeWolf was.

Yet some traditions endured, bridging the old and new eras. The quarterly video tape was still a staple, only now with Chamillard, Bradley, and Newman replacing d'Arbeloff and Van Veen. The TQM machinery was still in good working order, and huge TQM meetings (the 2000 edition was held at Boston's Fleet Center) stressed the importance the Company still placed on continuous improvement. Most important, the Company's management was almost entirely in the hands of people who had worked for Teradyne, and with d'Arbeloff, for many years. A few of them had even worked for Nick DeWolf. The culture of the Company was in good hands, and there would be no violent changes.

Yet there was a discernible change in management style as d'Arbeloff gave way to Chamillard. D'Arbeloff used his strong personality as a lever, and even where his arguments were not particularly strong they could be rammed home by a certain amount of constructive bullying. He was, moreover, a master motivator (some would say manipulator) of people, and he would tirelessly circulate among the troops, engaging them in debates on various points — debates he almost never lost.

Chamillard was more self-contained, preferring to challenge and debate his colleagues within the conference room, seldom roaming around the corridors in search of argument. Like d'Arbeloff, he had a good sense of humor, and like d'Arbeloff, he thrived on work. He was the prototypical manager, as d'Arbeloff was the prototypical entrepreneur. And just as the entrepreneur of 1960 had become the accomplished manager of the 90s, it would be up to the skilled manager who succeeded him to become enough of an entrepreneur to keep Teradyne charged with excitement.

There were few management changes in 1999, but those that occurred were significant. Dave Sulman and Ed Rogas were made Senior Vice Presidents on Chamillard's staff, while Brad Robbins, an ICD engineer (and Owen Robbins's son) took over management of the Division. Eileen Casal was hired as Teradyne's Vice President and General Counsel. And Randy Stone left her post as Service Manager to become Vice President for Information Systems. Steve Gladstone, managing Hammer Technologies, was also named a Teradyne Vice President.

The Board of Directors was expanded with the appointment of Roy Vallee, Chairman and CEO of electronics distributor Avnet.

24

• • •

Closing the Century with a Bang

Whether you believe that the century ended with the year 1999 or 2000 makes no difference. Either way, Teradyne closed it with a bang.

Only a couple of years before, when Teradyne's quarterly bookings at one point spiked up to $600 million, George Chamillard had used the quarterly video tape to warn employees not to take it too seriously. "Just because we booked $600 million in one quarter doesn't mean we're a $2.4-billion company," he said, no doubt worried that a surge in spending was in the making.

Yet here was Teradyne, in 2000, not a $2.4 billion-dollar company, but a $3 billion company. Teradyne had become, not just a billion-dollar company, but a company that could *grow* by a billion in a year. A Company with a billion-dollar *division* (TCS bookings topped that mark in 2000). Net income for the year topped a half billion dollars, for a dazzling 17-percent margin on sales. On Wall Street the champagne corks were popping as Teradyne shares soared through the $100-dollar mark. Once again, the year-to-year comparisons almost defied belief: growth of 140 percent at the Connection Systems Division, 70 percent in semiconductor test. Virtually everything Teradyne touched turned to gold. The Telecommunications Division successfully launched a new product called Celerity, designed to test the broad-band characteristics of telephone lines. Lest anyone miss the point, the unit rechristened itself, becoming Teradyne's Broadband Test Division. In circuit board test the focus shifted to optical inspection, an area Teradyne had entered a few years earlier by acquiring Control Automation. Now a new entrepreneurial group, born at MIT and brought to Teradyne by d'Arbeloff, was putting the final touches on a line of revolutionary new optical inspection systems.

Both strong-willed, opinionated people, d'Arbeloff and Chamillard could hardly be expected to see eye to eye on everything, but d'Arbeloff generally yielded to his successor, saying, "It's George's decision now" when asked about some matter of Company strategy. On the other hand, Teradyne's software test group was d'Arbeloff territory, and here it was Chamillard who stayed out of the way. In overseeing the software test group, d'Arbeloff also had by his side his two most trusted aides, Prestridge and Robbins, and of course he had a topnotch group manager in Jeff Hotchkiss.

By 1999 it had become obvious that two of the five operating units — Hammer and RSW — were successful, and the others were marginal. Within a year, Softbridge, Midnight, and Systems and Software Test would all be gone as separate entities, but valuable technology was extracted from each unit and added to the mix at Hammer, RSW, and the Telecommunications Division.

The prospect of a software-test spin-off, always on the table, now gathered momentum. The units themselves were ready to leave the nest. The business might, after all, grow faster on its own than it could under Teradyne's wing.

But the group was not ready for a public offering; that much Teradyne learned when it ran the idea by an investment banker. Its verdict: What Teradyne had was some interesting technology, some talented people, and some successful products. What it did not have, in any sense of the word, was a company. The software test group was, at best, a work in progress.

So attention turned to the possibility of finding another sponsor, one that could bring some valuable advice to the game. Teradyne, since it would in any event retain an interest, could only benefit if some other patron were to help fuel the growth.

D'Arbeloff, with his extensive knowledge of the venture capital community, came up with the new sponsor. Matrix Partners, one of the country's most successful venture firms, would serve as the lead investor in the new enterprise, with d'Arbeloff and other individuals also participating. Thus, at the end of the year 2000, majority ownership of the group, renamed Empirix, was transferred from Teradyne to its new owners.

The separation of Empirix from Teradyne also brought about the separation of d'Arbeloff from the Company he founded. The break was occasioned by the spin-off: D'Arbeloff the Empirix investor could not transact business with d'Arbeloff the Teradyne Board member. Those who knew d'Arbeloff knew that he would scrupulously avoid the reality of conflict, but there was no way to avoid the perception. He therefore left the Board and devoted his full energies to Empirix and to MIT, where he was still Chairman of the Trustees.

The year 2000 saw further fine-tuning of the new Teradyne management team. Joe Wrinn, who had come east to head the Company's combined board test operations, was brought into the Industrial/Consumer Division to help pilot Teradyne's largest ATE unit. Wrinn's slot as board-test manager was filled by Ron Butler. Butler had been Human Resources Manager but was hardly a stranger to board test, having run the Walnut Creek operation some years earlier. His HR duties were taken over by Jim Dawson, a seasoned HR manager and long-time Teradyne employee. And 2000 brought Teradyne yet another in-house counsel, Tom Grilk, replacing the departing Eileen Casal.

John Casey, a Vice President since 1990, was given responsibility for all semiconductor test field operations, including sales, applications, and service — a vast, worldwide army with a huge arsenal of resources at its disposal. Casey had long experience in Teradyne's semiconductor test area, interrupted by a short tour managing the Telecommunications Division.

Mark Jagiela, former Japan Division manager, had returned to the U.S. in 1999 as Engineering Manager for the VLSI Test Division in Agoura Hills. In 2000, Jagiela became VTD's Division Manager.

Pressed for capacity, TCS was expanding its facilities everywhere, but still barely keeping pace with demand. So it hit the acquisition trail in 2000, buying two closely related San Diego-area printed circuit board makers, Herco and Synthane-Taylor. Herco was known as a supplier of daughter boards, but it was sought by TCS for its potential as a maker of printed-circuit backplanes. (Synthane-Taylor was a Herco supplier.)

With the addition of Herco to its own expanded printed-circuit board facility in Nashua, the TCS product mix warranted the reorganization of the Division into two groups: Responsibility for "Components" (connectors and circuit boards) was given to Tom Pursch, and "Assembly" (electronic manufacturing services) was assigned to TCS veteran Wayne Morrison. Others reporting to Division Manager Schneider included Dan Murphy, running

Sales, Tom Pitten, heading Engineering, and Frank Carroccia, overseeing a broad range of finance, purchasing, facilities, personnel, and legal responsibilities.

Three new vice presidents were added to the roster in 2000: Jack Lanzoni, heading Purchasing; Marc Levine, the Integra Division Manager; and TCS's Tom Pursch.

In early 2000 the stock price topped $110, putting the Company's market value above $16 billion. As Teradyne followers have discovered time and again over the years, the market is a fickle mistress, and by year-end the stock, along with the stocks of many other technology companies, had lost 70 percent of its value. Those who insisted that "this time it's different" learned the hard way that, even if the underlying business is indeed different, the stock market is never different.

25

• • •

Retrospective

Alex d'Arbeloff's Teradyne years spanned the period from 1960 though 2000. From 1971 through 1997 he was the Company's Chief Executive Officer. In 1971, when he took over, the Company's sales were $13 million. The year he turned the Company over to George Chamillard, sales were $1.266 billion. But the sales figures, impressive though they are, miss the real story, which is d'Arbeloff's determination to build a company of real value, of enduring quality, of uncompromising integrity. The severe downturns of the 1970s, the LTX spin-off, the seemingly unstoppable rise of Far East competition — these were not allowed to derail d'Arbeloff from his goals. His success was a matter of will as much as of intellect and talent. But to say that is simply to restate one of life's truisms.

One can begin to understand the magnitude of d'Arbeloff's accomplishment only when one counts the bodies in the graveyards. Hundreds of companies were founded around 1960, along Route 128 and, later, along the Bay Shore Freeway. Some died in childbirth, some, like GCA and Computervision and Digital Equipment, became mighty companies, leaders in their industries. For each, somewhere along the line there was a fatal misstep, and that was that.

Teradyne under d'Arbeloff made its own missteps. In fact, a critic can cite a long list of d'Arbeloff-sponsored products that failed, some badly. But — and this is the point — they were never fatal, or even close to fatal. D'Arbeloff was willing to take risks, and he insisted that his people take risks, but he shrewdly controlled these risks and never hesitated to pull the plug on a venture going sour.

The team of senior managers d'Arbeloff assembled to help him run Teradyne is legendary. No ATE company, and few companies of any description, could match its lineup. D'Arbeloff also insisted that the Company hire the best and brightest college graduates it could find, in good years and bad, so that there would never be a shortage of promotable talent. It is no accident that most of today's officers and managers have many years of Teradyne service, and the college recruiting efforts are as intense as ever.

Looking back on the d'Arbeloff era, it would be unthinkable to omit the fact that it could not have happened if there had not been a DeWolf era. D'Arbeloff took the technology base created by his partner and built a great company on it. D'Arbeloff could not have built the base — that was not his talent — any more than DeWolf could have built a multibillion-dollar company. Those who would speculate on the relative value of their contributions might as well debate whether Jerome Kern or Oscar Hammerstein gave us "Old Man River."

26

• • •

Teradyne's Next Forty Years

Who would have dared dream, in 1970 or 1980, that Teradyne would become a $3 billion company? Not even Alex d'Arbeloff, who was ever the optimist, foresaw an ATE industry of its present scale. But then, who could have foreseen the Internet? Or gigabit memories or gigahertz processors? Or a display in your automobile's control panel that would talk you to your destination? Or music downloaded to your computer, or eBay or Amazon.com? The great thing about the technology of electronics is that invention begets invention that begets invention, and the third- and fourth-order effects are impossible to predict. Yet testing is a constant because, whatever the device, the good products and the bad ones look the same. Without testing, you are flying blind.

The same fundamental is at work in the connector business. Electrical signals and power sources must be interconnected to the real world, and the most complex, most artfully designed circuit is ultimately hostage to the integrity of its connectors.

Teradyne's future is sound because the Company is anchored to bedrock, not fads. Some years back, businesses placed their bets on whether digital watches would use LED or LCD readouts. They bet on the Beta or VHS videotape format, on laser disks versus DVD, on Apple versus PC, on wired versus wireless. Teradyne's game is different. It tries to cover the waterfront, so that, whoever wins, Teradyne wins.

Years ago, three prime contractors were bidding to supply equipment for a major U.S. military program called IFTI. To each of these contractors, the stakes were high. But not to Teradyne, whose digital test units were specified by all three of the bidders. Teradyne couldn't lose, and, with its present spectrum of ATE and connection products, the odds are the same: Teradyne can't lose. It will lose orders here and there, but over time it will grow to become a $5 billion company, a $10 billion company, and so on, limited only by the ingenuity of engineers designing new things to be tested.

Volatility is a given. The core business, the sale of semiconductor test systems, is a derivative of semiconductor sales, which is a derivative of electronics sales, which is a derivative of GDP. As you work your way down from macro- to microeconomics, the amplitudes of the cycles increase. That is fact, and "this time it's different" is fiction. Chip makers buy capital equipment based on increasingly optimistic sales forecasts, and eventually they have too much capacity and put on the brakes. As companies apply sophisticated computer programs to the problem, they become less vulnerable. But one component of the process is still human nature, where greed and fear often get in the way of common sense.

The volatility is not necessarily a bad thing, once you learn how to deal with it. After 40 years in the business, Teradyne has learned all the tricks. D'Arbeloff and Chamillard can smell an oncoming downturn as surely as a Maine fisherman can smell warm air coming from the south. The Company worries endlessly about its inventories and receivables, uses temporary help wherever possible, and subcontracts a growing percentage of its board manufacture. It husbands its backlog, generally setting the next quarter's shipments at the average of the preceding four quarter's bookings. When recession strikes and orders go into free-fall for a time, Teradyne reacts quickly and decisively to throttle back production but leaves engineering heavily funded. This gives it a clear advantage over smaller, financially weaker competitors struggling to survive.

The key to it all, of course, is the certain knowledge that you are in a long-term growth business, and therefore that a recovery is just a matter of time. That has always been a safe bet at Teradyne. If the day ever comes when the assumption is wrong, we are probably all doomed anyway.

As Teradyne begins its second 40 years, it is the undisputed king of the ATE business. But, this being a "what have you done for me lately?" business, no one at Teradyne is relaxing. With an annual engineering budget of a quarter billion dollars, the Company keeps rolling the new products out: a new Catalyst, called Tiger; a new AOI system, called Optima; a new memory tester, called Probe One; a new line test product, called Celerity.

Perhaps the most stunning transformation of the Teradyne landscape is the emergence of a new giant on the scene: Teradyne Connection Systems. In the year 2000, TCS booked more than $1 billion worth of orders. Exciting new products — the GbX high-density connector and the NexLev stacking connector — were on the way. A major new TCS plant was on line in Ireland, and operations in the U.S. were being expanded rapidly. In Japan, Rick Dyck, who managed Teradyne's affairs in that country until 1997, is now the proprietor of his own business — bringing TCS products to the Japanese market. The Dyck enterprise is called TCSJ, and it is very successful.

Meanwhile, back in Harrison Avenue, George Chamillard runs the Company from his tenth-floor cubicle. It is doorless, in the Teradyne tradition, and it is quite small, also in the Teradyne tradition. Next door (if there were a door) sits CFO Mike Bradley, and nearby are John Casey, Dave Sulman, and Roger Saylor. Ed Rogas is usually near at hand, though he calls California home. The division managers, Felding for memory, Jagiela for VLSI, Lasson at Broadband, Schneider at TCS, Butler at board test, are an e-mail or a phone call away, as are Osattin in Finance, Dias in Operations, and all the others. Good people, also in the best Teradyne tradition.

Epilogue

Teradyne's history is obviously a work in progress, and we leave it to some future historian to pick up the trail from the beginning of 2001 on. That story will begin with Teradyne facing another sharp business downturn. The feast that the semiconductor and semiconductor capital equipment industries enjoyed for the last several years of the century ended abruptly as the year 2000 ended, and orders were falling sharply across most of Teradyne's product lines. Having been there more than once in the past, Teradyne's management is doing what it must to insulate the Company from any long-term damage, and if there is any lesson in the story we have told to date it is that the Company will see its sales soar to new heights, once the downturn has run its course.

Some readers may wonder, in reading about the people who made Teradyne a great company, where they all are now. The founders are both as active and enthusiastic about their ventures as they were back in 1960. Nick and Maggie DeWolf live near the center of Aspen, and he is deeply involved in a great many projects, recording Vietnamese music, sponsoring a science lecture series, running Web sites, and generally shaking up the world around him. Alex and Brit d'Arbeloff still live in Brookline, and he splits his time between MIT, where he is Chairman of the Trustees, and a number of business ventures, including the Teradyne spin-off Empirix.

A number of very special Teradyne people, including Abe Bluestone, Milt Collins, Stu Peters, Pete Sylvan, Walter Kern, and Warren Cozzens have died. Russell Ashdown's company was bought by Texas Instruments, and he is running it for them. Chuck Pearson also runs a small high-tech company, and Byron Brooks works with him. Dennis O'Connell lives in his beloved England with his wife Violette, and the two travel extensively. Jim Prestridge divides his time between homes in Arizona and Wyoming, Owen Robbins still calls the Boston area home, and both are on the Board of Empirix. The three TCS founders, Windy Hunter, Len Johnson, and Tom Neavitt, are all enjoying retirement, as are Joe Hendriks and Dick Lucey.

APPENDIX I

The Teradyne Business Plan

I. General Objective

It is the objective of Teradyne to design, manufacture, and market electronic equipment for industrial use. This equipment will be designed to be free from failure for five years. Its design will also aim at a minimum period between required calibrations and adjustments to be in excess of three months under the rigorous conditions that changes in temperature, variances in power line supply, and other environmental factors produce. These capabilities will exceed those of most presently available equipment by a ratio of over ten to one. In addition, basic accuracy will be improved several times.

Electronic equipment for industry requires a high order of reliability and dependability. The high cost of down-time for maintenance and repeated calibration is sufficient to warrant more care in the design of such equipment and the use therein of higher quality components than in the production of consumer goods. The penalties to the user of undetected improperly functioning equipment may be many times the original cost of the equipment.

A new capability for producing dependable electronic equipment has arisen and become economically feasible within the last few years. Achievement of high reliability in such equipment and its components, which has been demanded by the military, has now become practical for the industrial market. However, the organization of companies engaged in military electronics has made their penetration into industry very difficult.

The experience of the Company's management is particularly suited to this field. The economic and competitive application of high reliability principles requires careful planning and competence in all phases of the Company's operation from basic design through manufacture, supplier selection, field installation, and service.

II. Modular Design

In most of the branches of industrial electronics, a high proportion of equipment is designed from the ground up, an economically unsound practice. A more reasonable approach is to build modules or building blocks which can then be assembled in various configurations. However, considerable judgment and experience are necessary to make this approach successful. Our intention is to utilize a well-balanced series of small modules which will permit the construction of a very large portion of our equipment with an absolute minimum of design and application engineering. This modular system will permit large improvements in manufacturing efficiency and also, through total parts limitation, in inventory control.

III. Development

The management of Teradyne has excellent contacts and a high reputation in the electronics industry. However, the Company has no intention of going to the market until it has a flexible line of quality products and can assure reasonably short delivery on orders received. The Company's development stage will be used to perfect modules which combine maximum flexibility with optimum economic utilization, both electronic and mechanical. Many novel and proprietary techniques are anticipated in the design of these universal building blocks.

Although the primary effort in this stage will be devoted to development of the modular concept, the Company is already developing a specific testing instrument for which enthusiastic market acceptance is anticipated. This instrument will be constructed to serve as a demonstrator of Teradyne's capabilities. It can be modified easily to serve a very broad range of requirements.

IV. Manufacturing

The growth of the Company will be paced by the rate at which a sound, integrated, and efficient manufacturing effort can be established. Because of the high quality nature of our products, careful training and supervision are necessary, and a carefully planned build-up rate will be established. Manufacturing methods will be simplified by the design techniques employed. Inventory control is considered of prime importance and will be simplified by engineering discipline and a very limited approved standard parts system.

V. Market and Competition

The initial market that Teradyne will enter is the field of instrumentation and automation for manufacturers and users of semiconductors and other electronic components. The semiconductor grading method in which the sales price of a given device is determined by a complex testing procedure emphasizes the cost penalties of inadequate or unreliable equipment. Although conventional laboratory instruments are often used, they are not satisfactory for this service, which demands a high reliability approach. The management of Teradyne is particularly familiar with and well recognized in this field.

The lack of competent and capable producers in the component testing field has forced many users to construct their own equipment. Because this effort is a distraction from their main effort, there is a definite trend toward the purchase of equipment from outside. Many new and small companies have been formed to exploit this large market, but none have clearly risen to prominence or established a definite reputation. It is our intention to become a strongly sales oriented company. Our contacts and sales capability are vital to our success.

VI. Business Outlook

A hard selling approach coupled with accurate pricing should result in good and consistent profits. The large market enables us to limit our efforts to areas of our immediate capability. New products and modules will be developed on the Research and Development budget where their priority in terms of Company growth warrants it.

We believe that our approach in a large and growing market will permit Teradyne to rise to a key and well-recognized position and to build a strong and capable organization.

We believe it would be a mistake for our Company to rely too strongly in the early stages on a fixed product in a narrow technical area. It is anticipated, however, that the manufacturing and marketing organization built up over the first several years of operation will place Teradyne in a sound position for entering the fixed product area when it is ready to do so. Industrial electronics has a very broad spectrum of need for such products of high reliability.

APPENDIX II

Teradyne's Basic Strategies for Success in the Test Equipment Business

("The Green Book")
Published 1974

Introduction

The following is a summary of the basic strategies underlying Teradyne's past and present success in the test-systems business. We set them down here not as articles of faith to endure through the ages, but simply as an indication of what Teradyne stands for today. We cannot know where we are going unless we first know where we are, and if this document serves as nothing more than a starting point for discussion, it will have served a useful purpose.

A statement of principles can also serve as an operating guide in certain situations. Usually, plain good sense will lead people to make correct decisions without benefit of guidelines. It is at the borderline, where a decision may go either way, that explicitly stated principles are most helpful. Since such borderline decisions are often the critical decisions on which a company's fortunes turn, it is essential that the underlying principles be clearly understood by all.

Much of what follows has to do with the motivation of creative people. Because Teradyne is in a highly technical business, a major part of our creative resources are technical. Nevertheless, it would be a mistake to conclude that creativity at Teradyne is the exclusive property of our engineers and technicians. We have creative people in every corner of our Company, and we mean the following to apply to all of them.

Finally, we do not claim that the principles set forth here are the only ones that work or that we have a monopoly on business wisdom. Nor do we mean to apply these principles blindly to every operation that bears the Teradyne name. For our core business, however — the design, production, and sale of automatic test systems — we think that they are principles that make sense in today's world as we find it.

Teradyne's Over-all Objective...

... is to offer its customers, employees, and stockholders a balanced program of growth based on profitability.

To accomplish this objective, two attributes that have characterized its past must be used to shape its future. They are:

1. A fundamental understanding of technical problem-solving applied to automation within the electronics industry.
2. An organization that serves as a framework for individual creativity tempered by common goals.

How Teradyne Serves its Customers, Employees, and Stockholders

To be successful, Teradyne must serve its customers, its employees, and its stockholders. The needs of these three groups are interdependent, and no one of the groups can long be served at the expense of the other two. The stockholder must be assured that his investment in Teradyne offers a promise of increasing earnings per share. The employee must be assured that he will share in whatever success he helps the company achieve. The customer must be assured that the company's products and services will provide economical, practical solutions to his problems. These various interests can be served only if the company is profitable. In a growing industry, moreover, continuity of profits can be ensured only by company growth. Thus Teradyne's single overriding objective must be *profitable growth*.

At Teradyne, we have achieved profitable growth by applying individual creativity to projects chosen for their potential contribution to the Company's over-all development.

Applying and Coordinating Individual Creativity

Because we must coordinate a variety of creative talents — both technical and nontechnical — in order to serve our customers' problems, it is absolutely essential that the objectives of the individuals who comprise Teradyne are in harmony with the over-all company objective. A creative individual's objectives are likely to center on freedom of action, a sense of personal involvement in an interesting task, and the sharing of any success he helps to bring about. We ensure his freedom of action by minimizing the supervisory and supporting staffs surrounding him. We heighten his sense of personal involvement by bringing him into the most direct contact with our customers' problems. And we reward the creative individual who works best alone no less than we reward the person who works through others.

Project Selection

Teradyne's business is based on the art of doing simple things well. We are not interesting in technological achievement for its own sake, but only as it directly and immediately offers better solutions to real problems. We define real problems in terms of our customers' needs rather than their wants. The suggestion that we know better than our customers what these needs are is neither presumptuous nor arrogant; customers quite understandably want improved versions of what they already know, whereas Teradyne's contribution lies in innovation.

Each project that Teradyne undertakes should lay the foundation for the next. Therefore, the human and financial resources committed to any project must not only generate the answers to the customer's specific problems, but must add to the skills and competence that will be available for future customers. A project that does not contribute to the systematic body of knowledge necessary for technical problem-solving is at best a digression and at worst a true cost to the Company, representing opportunities foregone. Since Teradyne provides its customers not simply with hardware but with solutions to problems, we must, when undertaking a project, know as much about the problem involved as the customer does. That means that we must commit to each problem a considerable measure of our intellectual resources — our Company's most valuable asset. If the project falters, further resources will have to be diverted away from other problems, and the orderly development of our total range of competence will be disrupted. Before accepting a project, therefore, we must thoroughly understand the risks associated with it. This does not mean that Teradyne will not take risks. It does mean that we expect all concerned to be aware of the magnitude of the risks associated with each project and appreciate the far-reaching consequences of a bad gamble.

With an organization designed to serve the needs of our customers, we offer as our only "standard" product a fundamental understanding of technical problem-solving, especially as it relates to automation in the electronics industry. As we grow, our broad experience in this general area will expose us to a great many opportunities. We will evaluate these various opportunities in terms of three criteria: (1) our ability to learn the customer's problem; (2) the size of the potential market relative to our Company's size; and (3) our chances of obtaining a dominant share of the market.

Summary

1. Our over-all objective is profitable growth.
2. We achieve profitable growth by applying well motivated individual creativity to the solution of customers' problems.
3. We motivate creative individuals by giving them freedom of action, personal involvement in interesting projects, and a just share of the Company's success.
4. Creative individuals should be allowed to operate with a minimum of supervision and support.
5. An individual who works best alone should share in the Company's success no less than one who works through others.
6. Teradyne's business centers on the art of doing simple things well.
7. Teradyne responds to customers' needs, which may or may not be the same as their wants.
8. Each project undertaken should offer a practical solution to a real problem and should lay the foundation for a subsequent project.
9. We must, in undertaking a project, know as much about the customer's problem as the customer does.
10. Our success in the test-equipment field requires a dominant share of a well-defined market.

APPENDIX III

Annual Sales, as Reported ($M)

1961	0.005	1981	159.942
1962	0.156	1982	176.305
1963	0.431	1983	251.422
1964	0.913	1984	389.278
1965	1.902	1985	336.368
1966	3.81	1986	306.112
1967	5.581	1987	377.658
1968	9.112	1988	462.297
1969	15.383	1989	483.575
1970	18.172	1990	458.877
1971	13.08	1991	508.923
1972	20.661	1992	529.581
1973	40.66	1993	554.734
1974	56.203	1994	677.44
1975	37.872	1995	1191
1976	53.731	1996	1171.6
1977	67.815	1997	1266.3
1978	91.663	1998	1489.2
1979	122.447	1999	1790.9
1980	164.97	2000	3044.3